Revise AS Chemistry for Salters (OCR)

Ann Dani... ...on and Chris Otter

www.heinemann.co.uk

✓ Free online support
✓ Useful weblinks
✓ 24 hour online ordering

01865 888058

Heinemann

Inspiring generations

Heinemann Educational Publishers
Halley Court, Jordan Hill, Oxford OX2 8EJ
Part of Harcourt Education

Heinemann is the registered trademark of
Harcourt Education Limited

© Ann Daniels, Lesley Johnston, University of York Science Education Group 2004

First published 2004

08 07 06
10 9 8 7 6 5 4

British Library Cataloguing in Publication Data is available
from the British Library on request.

10-digit ISBN : 0 435583 46 8
13-digit ISBN: 978 0 435583 46 0

Edited by Tim Jackson

Index compiled by Ann Hall

Designed and typeset by Saxon Graphics Ltd, Derby

Original illustrations © Harcourt Education Limited 2004

Printed and bound in Great Britain by Ashford Colour Press Ltd, Gosport, Hants.

Acknowledgements
The examination questions and mark schemes are reproduced by kind permission of OCR and
are taken from the following papers:

Page 74, Q1: June 2003
Page 75, Q2: June 2003
Page 76, Q3: June 2002
Page 77, Q4: June 2000
Page 78, Q5: January 2003

Every effort has been made to contact copyright holders of material reproduced in this book.
Any omissions will be rectified in subsequent printings if notice is given to the publishers.

Contents

Introduction – How to use this revision guide

This revision guide is for the OCR Chemistry (Salters) AS course, and is valid for examinations taken from June 2004 onwards. You may take examinations in January or June, or just in June at the end of the course.

This guide covers the two written examinations for the AS course, **Module 2850, Chemistry for Life** and **Module 2848, Chemistry of Natural Resources**.

The table below shows the scheme of assessment for the AS course. This enables you to see how the examination modules and teaching content link together.

Examination Module	Module title and teaching topics covered	Duration	Number of marks	Mode of assessment	% weighting of AS
2850	**Chemistry for Life** (The Elements of Life and Developing Fuels)	1hr 15min	75	Written examination	30%
2848	**Chemistry of Natural Resources** (From Minerals to Elements, The Atmosphere and The Polymer Revolution)	1hr 30min	90	Written examination	40%
2852	**Skills for Chemistry**	Not covered in this revision guide – your school or college will organise this assessment	Coursework	30%

If you choose to go on to the second year of the course the marks gained in the AS year will count as 50% of your final GCE A level marks.

At the start of each teaching topic in this book there is an introduction which summarises the content. It allows you to see which sections from **Chemical Ideas** and **Chemical Storylines** need to be revised for that specific topic.

Within each teaching topic the material is divided into short sections of one to three pages. Each section usually revises one section of material from Chemical Ideas, and ends with **quick check questions** to help you test your understanding.

Also included in the book are sections on **experimental techniques** and **examination hints and tips**.

Towards the end of the book there are **practice examination questions**, along with helpful comments to assist in answering.

All questions (quick check and examination questions) are provided with full answers.

The Elements of Life (EL)

This unit tells the story of the elements of life: what they are, how they originated and how they can be detected and measured. As you work through this unit you will cover the following concepts. CI refers to sections in your Chemical Ideas textbook.

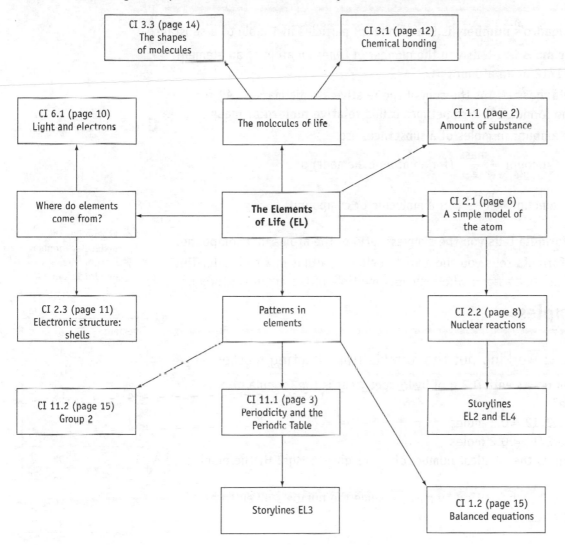

Amount of substance
Chemical Ideas 1.1

What you need to know

- 6.02×10^{23} (**Avagadro's number**) is the number of particles in 1 mole of a substance.
- **Relative atomic mass** (A_r) tells you the number of times an atom of an element is heavier than 1/12 of an atom of ^{12}C.
- **Relative formula mass** (M_r) is the sum of the relative atomic masses (A_r) for each atom in the formula. It is sometimes called **relative molecular mass**.
- To work out the amount in moles of a substance, use

$$\text{amount} = \frac{\text{mass}}{A_r} \text{ (for an atomic element) or}$$

$$\text{amount} = \frac{\text{mass}}{M_r} \text{ (for a molecule or compound)}$$

- The **empirical formula** tells you the <u>simplest ratio</u> of the atoms in a compound.
- The **molecular formula** tells you the <u>actual number</u> of atoms in a molecule. This may be either the same as or a whole number multiple of the empirical formula.

▶ CO_2: 12 + 16 + 16 = 44

▶ Ethane has the molecular formula C_2H_6 but its empirical formula is CH_3.

Worked examples

Worked example 1: Working out the formula from reacting masses

1.2 g of magnesium reacts with 0.2 g of hydrogen. What is the formula of magnesium hydride?
- 1.2 g of Mg = 1.2/12 = 0.1 moles
- 0.2 g of H = 0.2/1 = 0.2 moles
- Dividing through by the smallest number of moles gives 1 Mg:2 H. The empirical formulae is MgH_2.

If the question gives % composition by mass, assume the numbers given to be grams and proceed as above.

element	A_r
C	12
O	16
Mg	24
H	1
N	14
S	32
Na	23
Pb	207
Be	9

A_r values will always be given to you in a question.

Worked example 2: Working out the percentage by mass of an element, given the formula of the compound

Calculate the percentage by mass of nitrogen in ammonium sulphate, $(NH_4)_2SO_4$.

Step 1 $M_r = 2 \times (14 + 1 + 1 + 1 + 1) + 32 + 4 \times 16 = 132$

Step 2 mass of nitrogen in 1 mole of $(NH_4)_2SO_4$ = 14 + 14 = 28

Step 3 percentage by mass = $\dfrac{\text{mass of element in 1 mole of the compound}}{M_r} \times 100 = \dfrac{28}{132} \times 100 = 21.2\%$

❓ Quick check questions

1 Work out the relative formula masses of: (a) MgO; (b) Na_2SO_4; (c) $Pb(C_2H_5)_4$.

2 A compound is 12.9% Be, 17.3% C and 69.8% O by mass. Calculate its formula.

3 Calculate the percentage by mass of lead in $Pb(C_2H_5)_4$.

Periodicity and the Periodic Table

Chemical Ideas 11.1 and 2.3 and Chemical Storylines EL3

The Periodic Table of the Elements

1	2											3	4	5	6	7	0
				1.0 H Hydrogen 1													4.0 He helium 2
6.9 Li lithium 3	9.0 Be beryllium 4											13.8 B boron 5	12.0 C carbon 6	14.0 N nitrogen 7	16.0 O oxygen 8	19.0 F fluorine 9	20.2 Ne neon 10
23.0 Na sodium 11	24.3 Mg magnesium 12											27.0 Al aluminium 13	28.1 Si silicon 14	31.0 P phosphorus 15	32.1 S sulphur 16	35.5 Cl chlorine 17	39.9 Ar argon 18
39.1 K potassium 19	40.1 Ca calcium 20	45.0 Sc scandium 21	47.9 Ti titanium 22	50.9 V vanadium 23	52.0 Cr chromium 24	54.9 Mn manganese 25	55.8 Fe iron 26	58.9 Co cobalt 27	58.7 Ni nickel 28	63.5 Cu copper 29	65.4 Zn zinc 30	69.7 Ga gallium 31	72.5 Ge germanium 32	74.9 As arsenic 33	79.0 Se selenium 34	79.9 Br bromine 35	83.8 Kr krypton 36
85.5 Rb rubidium 37	87.6 Sr strontium 38	88.9 Y yttrium 39	91.2 Zr zirconium 40	92.9 Nb niobium 41	95.9 Mo molybdenum 42	Tc technetium 43	101 Ru ruthenium 44	103 Rh rhodium 45	106 Pd palladium 46	108 Ag silver 47	112 Cd cadmium 48	115 In indium 49	119 Sn tin 50	122 Sb antimony 51	128 Te tellurium 52	127 I iodine 53	131 Xe xenon 54
133 Cs caesium 55	137 Ba barium 56	139 La lanthanum * 57	178 Hf hafnium 72	181 Ta tantalum 73	184 W tungsten 74	186 Re rhenium 75	190 Os osmium 76	192 Ir iridium 77	195 Pt platinum 78	197 Au gold 79	201 Hg mercury 80	204 Tl thallium 81	207 Pb lead 82	209 Bi bismuth 83	Po polonium 84	At astatine 85	Rn radon 86
Fr francium 87	Ra radium 88	Ac actinium ** 89	Rf rutherfordium 104	Db dubnium 105	Sg seaborgium 106	Bh bohrium 107	Hs hassium 108	Mt meitnerium 109	Unn ununnilium 110	Uuu unununium 111	Uub ununbium 112		Uuq ununquadium 114		Uuh ununhexium 116		Uuo ununoctium 118

lanthanides *

140 Ce cerium 58	141 Pr praseodymium 59	144 Nd neodymium 60	Pm promethium 61	150 Sm samarium 62	152 Eu europium 63	157 Gd gadolinium 64	159 Tb terbium 65	162 Dy dysprosium 66	165 Ho holmium 67	167 Er erbium 68	169 Tm thulium 69	173 Yb ytterbium 70	175 Lu lutetium 71

actinides **

Th thorium 90	Pa protactinium 91	U uranium 92	Np neptunium 93	Pu plutonium 94	Am americium 95	Cm curium 96	Bk berkelium 97	Cf californium 98	Es einsteinium 99	Fm fermium 100	Md mendelevium 101	No nobelium 102	Lr lawrencium 103

metals

non-metals

key

6.9 — relative atomic mass
Li — symbol
lithium
atomic number — 3

Mendeleev and the Periodic Table

- Mendeleev arranged elements in order of **relative atomic mass**.
- He then swapped elements over if he thought that they fitted better into another group based on their physical and chemical properties.
- He left <u>gaps</u> for elements which he correctly thought were yet to be discovered.

We now arrange the elements in order of <u>atomic number</u> rather than relative atomic mass.

> Mendeleev arranged elements in order of <u>relative atomic mass</u>, not mass number or atomic number.

Periodicity

Periodicity is exhibited when:

- There is a <u>regular pattern</u> in a property as you go across a Period.
- The regular pattern is <u>repeated</u> in other Periods.

Examples of periodicity

1. Melting point and boiling point

The pattern is for the melting point to increase and then decrease across a Period. This pattern is repeated in more than one Period. There is periodicity in melting points.

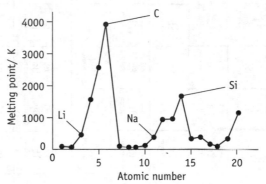

> On the left of the Period, metallic bonding is strong and a lot of energy is required to overcome the electrostatic attractions between the metallic ions and the delocalised electrons. On the right of the Period, weak intermolecular forces do not require much energy to break them.

2. Electrical conductivity

The pattern is for the electrical conductivity to increase and then decrease across a Period. This pattern is repeated in more than one Period. There is periodicity in electrical conductivity.

> Metals are good conductors of electricity due to the mobility of delocalised electrons. Non-metals are usually poor conductors of electricity since the electrons are held in covalent bonds.

3. Ionisation enthalpies

The first ionisation enthalpy is the energy required to <u>remove an electron</u> from <u>one mole of atoms</u> of the element in the <u>gaseous state</u>. The units are kJ mol^{-1}.

$$X(g) \rightarrow X^+(g) + e^-$$

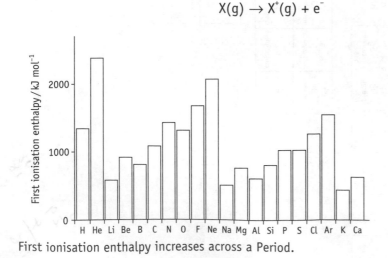

First ionisation enthalpy increases across a Period.

> Remember to put (g) in ionisation enthalpy equations.

> 1st ionisation enthalpy increases across a Period.

Using the Period 2 elements (Li to F) as an example, the electron to be lost is always from the second electron shell. However, the nuclear charge increases on going across the Period. This means that there is a greater attraction between the nucleus and the electron and more energy is needed to remove it. For example, the first ionisation enthalpy for fluorine is greater than that for oxygen. Fluorine is very unlikely to form F$^+$.

When they react, sodium (2.8.1) and potassium (2.8.8.1) both lose the one outer shell electron to form positive ions. Potassium is more reactive than sodium because the outer shell electron is further from the nucleus, is less firmly held and more easily lost. The first ionisation enthalpy for potassium is therefore lower than that for sodium. Potassium is more likely to form K^+ than sodium is to form Na^+ since less energy is required.

> 1st ionisation enthalpy decreases down a Group.

The first ionisation energy increases across Period 2 and this pattern is repeated across Period 3. Periodicity is exhibited.

Successive ionisation enthalpies

$$\text{1st:} \quad X(g) \rightarrow X^+(g) + e^- \qquad \text{3rd:} \quad X^{2+}(g) \rightarrow X^{3+}(g) + e^-$$
$$\text{2nd:} \quad X^+(g) \rightarrow X^{2+}(g) + e^- \qquad \text{4th:} \quad X^{3+}(g) \rightarrow X^{4+}(g) + e^-$$

Successive ionisation enthalpies increase for a particular element.

i.e. 4th > 3rd > 2nd > 1st

There is an increase in successive ionisation energies for a particular element. Moving to the next shell in to remove an electron requires much more energy since

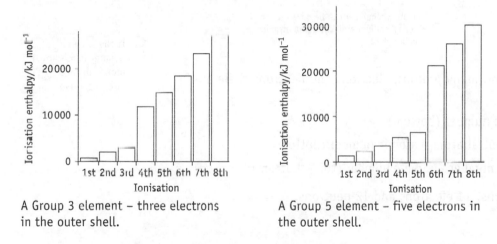

A Group 3 element – three electrons in the outer shell.

A Group 5 element – five electrons in the outer shell.

that electron is closer to the nucleus.

? Quick check questions

1. How does the melting point of the elements in the 3rd Period change on going from left to right?

2. The first ionisation energies for S and Se are +1000 kJ mol^{-1} and +940 kJ mol^{-1}. Predict the first ionisation energy for Te.

3. Suggest reasons why the first ionisation enthalpy of calcium is larger than the first ionisation enthalpy of potassium and why the first ionisation enthalpy of potassium is smaller than the first ionisation enthalpy of sodium.

4. Write an equation for the second ionisation enthalpy of calcium.

A simple model of the atom

Chemical Ideas 2.1

Inside the atom

The three types of sub-atomic particles have different masses and charges:

Particle	Mass on relative atomic mass scale	Charge
Proton	1	+1
Neutron	1	0
Electron	very small (0.00055)	−1

Mass number
= protons and neutrons

Atomic number
= protons

$^{23}_{11}\text{Na}$

^{23}Na

The atomic number can be omitted since all sodium atoms have an atomic number of 11

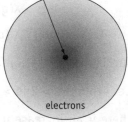
protons and neutrons in nucleus

electrons

Simple model of an atom.

The nuclear symbol tells you how many protons, electrons and neutrons there are in an atom of an element.

- Number of protons = atomic number (bottom).

- Number of electrons in a neutral atom = atomic number (bottom).

- Number of neutrons = mass number − atomic number (top − bottom).

So a sodium atom has 11 protons, 11 electrons and 12 neutrons.

> ◖ If the atomic number is missing, look it up in a Periodic Table.

Isotopes

Isotopes are atoms of the same element. They have the <u>same atomic number</u> (number of protons), or they wouldn't be the same element, but they have <u>different mass numbers</u> (and so have different numbers of neutrons).

For example:

$^{35}_{17}\text{Cl}$ has 17 protons and 18 neutrons

$^{37}_{17}\text{Cl}$ has 17 protons and 20 neutrons

Pure samples of the isotopes chlorine-35 and chlorine-37 would have different relative atomic masses (or different relative isotopic masses). Chlorine gas is a mixture of the two isotopes. Knowing the <u>relative isotopic masses</u> and the <u>relative abundance</u> of the two isotopes allows you to calculate the **relative atomic mass** of the mixture. The relative atomic mass of the mixture is the average of the relative isotopic masses.

If there is 75% of ^{35}Cl and 25% of ^{37}Cl:

$$\text{relative atomic mass} = \frac{(35 \times 75) + (37 \times 25)}{100} = 35.5$$

> ◖ If you are asked to explain what the term 'isotopes' means using chlorine-35 and chlorine-37 then, after the definition, give the numbers of protons and neutrons in each isotope and say what is the same and what is different.

> ◖ If you are asked to quote the relative atomic mass to a certain numbr of decimal places, then do so.

The mass spectrometer

The mass spectrometer can tell us about the relative isotopic masses and relative abundance of isotopes.

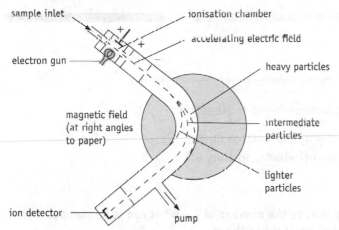

sample inlet: gas or liquid injected
electron gun: heated filament produces high-energy electrons
ionisation chamber: high-energy electrons bombard atoms of the sample and knock electrons out – positive ions of sample are formed
accelerating electric field: this speeds up the positive ions of the sample
magnetic field: this deflects or bends the stream of positive ions – lighter ions are deflected most
ion detector: this detects the ions – by altering the magnetic field, all the isotopes can be made to hit the detector

You may be asked to label a mass spectrometer or to describe what happens inside the mass spectrometer.

A mass spectrum for chlorine is shown below.

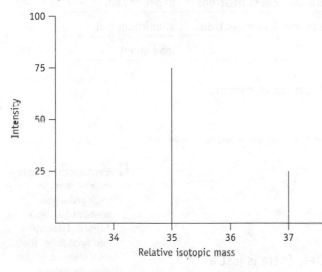

Relative isotopic mass	Relative abundance, %
35	75
37	25

You may be asked to sketch a mass spectrum using data from a mass spectrometer.

Quick check questions

1 How many protons, neutrons and electrons are there in (a) ^3_1H; (b) ^{47}Ca; (c) tellurium-122; (d) americium 241; (e) $^{16}_8O^{2-}$?

2 Calculate the relative atomic mass of magnesium from this data. Give your answer to 1 decimal place.

Mass number of isotope	Relative abundance, %
24	70
25	19
26	11

3 Sketch the mass spectrum produced by the sample of magnesium in question 2.

Nuclear reactions
Chemical Ideas 2.2, Chemical Storylines EL2 and EL4

Radioactive isotopes

Some isotopes have **unstable nuclei**, this results in the isotope being **radioactive**.

Three types of rays/particles can be emitted spontaneously by radioactive nuclei. All three types of emissions are capable of knocking electrons off atoms, so they are sometimes referred to as **ionising radiation**.

Radiation	What is it?	What happens to the number of protons and neutrons in the nucleus after emission?	What can it be stopped by?
α	a stream of helium nuclei, $^{4}_{2}He$	2 less protons and 2 less neutrons	paper or skin
β	a stream of electrons, $^{0}_{-1}e$	1 more proton and 1 less neutron	aluminium foil
γ	electromagnetic radiation	no change	lead sheet

All three types of emissions are dangerous to humans and can cause cancer.

Equations for radioactive decay

- α particles   e.g. $^{238}_{92}U \rightarrow {}^{234}_{90}Th + {}^{4}_{2}He$

  > Because the atomic number has changed a new element has been formed. Look up the symbol for the new element in the Periodic Table.

- β particle   e.g. $^{14}_{6}C \rightarrow {}^{14}_{7}N + {}^{0}_{-1}e$

- γ radiation   There is no change in the atomic particles. There is just a release of energy from the nucleus. γ radiation often accompanies the emission of α and/or β particles.

The nuclear equations are often a little tricky so read the question carefully.

> The sum of the mass numbers on the left of the arrow in the equation always adds up to the sum of the mass numbers on the right of the equation arrow. The same applies to the atomic numbers.

Example

Beryllium-9 can be bombarded with α particles. This causes neutrons to be released and carbon-12 is produced.

$$^{9}_{4}Be \quad + \quad ^{4}_{2}He \quad \rightarrow \quad ^{12}_{6}C \quad + \quad ^{1}_{0}n$$

$^{9}_{4}Be$	$^{4}_{2}He$	$^{12}_{6}C$	$^{1}_{0}n$
look up the atomic number in the Periodic Table	'bombarded with'	6 is the atomic number for carbon	'released'

> an α particle is $^{4}_{2}He$ or $^{4}_{2}\alpha$
> a β particle is $^{0}_{-1}e$ or $^{0}_{-1}\beta$
> a neutron is $^{1}_{0}n$
> a proton is $^{1}_{1}p$

Half-lives

The half-life is the time it takes for half of the nuclei in a radioactive isotope to decay.

Worked example

Iodine-131 has a half-life of 8 days. What fraction of the sample remains after 24 days?

Step 1 After 8 days, $\frac{1}{2}$ of the sample remains.

Step 2 After 16 days, $\frac{1}{4}$ of the sample remains.

Step 3 After 24 days, $\frac{1}{8}$ of the sample remains.

Tracers

Tracers are radioactive isotopes which can be used in medicine to aid diagnosis. The tracer is eaten, injected or drunk and then its pathway is followed using a **Geiger counter**. Isotopes emitting β radiation are usually chosen since they aren't stopped by skin. An isotope should have a half-life which is neither too short (or it will decay before tracing is complete) nor too long (or it will persist for too long in the body, potentially causing harm to the patient). For details, see Chemical Storylines EL2.

Nuclear fusion

Nuclear fusion is the joining together of two nuclei to form a heavier nucleus of a new element. High temperatures are required to provide the energy needed to overcome the repulsion between two positive nuclei. Nuclear fusion occurs during star formation. Examples are:

$$^1_1\text{H} + {}^2_1\text{H} \rightarrow {}^3_2\text{He}$$

$$3\,{}^4_2\text{He} \rightarrow {}^{12}_6\text{C}$$

For details, see Chemical Storylines EL4.

Quick check questions

1 Complete these nuclear equations:

(a) $^{47}_{20}\text{Ca} \rightarrow {}^0_{-1}\text{e} + \underline{{}^{47}_{21}}$

(b) $^{226}\text{Ra} \rightarrow {}^4_2\text{He} + \underline{{}^{222}}$

(c) $^{241}_{95}\text{Am} \rightarrow {}^4_2\text{He} + \underline{{}^{237}_{93}}$

(d) $^{131}\text{I} \rightarrow {}^0_{-1}\text{e} + \underline{\quad}$

2 An antique watch has hands painted with radioactive paint. Would it affect the wrist of the wearer?

3 $^{47}_{20}\text{Ca}$ has a half-life of 6.5 days. What fraction of the sample will be left after 26 days?

4 Name the equipment used to detect β particles.

Light and electrons
Chemical Ideas 6.1 and 2.3

Energy levels

An electron in a hydrogen atom can occupy any one of the **fixed energy levels**. These energy levels are the same for all hydrogen atoms. In the ground state, the electrons are closest to the nucleus and have the least energy. Notice that the difference in the energy levels (ΔE) decreases as the electron moves away from the nucleus.

Electrons always move to specific levels, never in between. Therefore, if you are asked to show an electron transition on a diagram always draw from one horizontal line to another – never start or finish between lines.

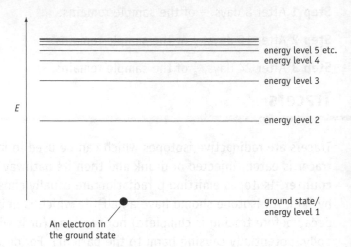

Absorption spectra

These are seen from Earth when atoms in the chromosphere around the Sun absorb light.
● Electrons absorb a 'photon' or package of energy.
● Excited electrons move up to a higher energy level – they are promoted.

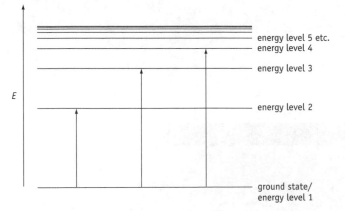

● The electromagnetic radiation absorbed by each of the hydrogen atoms has a definite frequency related to the difference in energy levels by $\Delta E = h\nu$.

An absorption spectrum seen on Earth is the spectrum of visible light (which looks like a rainbow) with black lines corresponding to the absorptions of energy by the electrons.

An absorption spectrum.

Emission spectra

These are seen when a chemical burns with a coloured flame.
● Electrons absorb a 'photon' or package of energy.
● Excited electrons move up to a higher energy level – they are promoted.
● Electrons then drop back to lower energy levels.

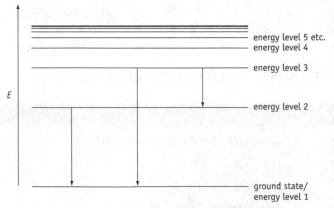

● The electromagnetic radiation emitted by each of the hydrogen atoms has a definite frequency related to the difference in energy levels by $\Delta E = h\nu$.

An emission spectrum has a black background with coloured lines on it. These coloured lines correspond to the emissions of energy by the electrons.

An emission spectrum.

Electron shells

Energy levels are usually referred to as **electron shells** for atoms more complex than hydrogen. Each shell can hold a certain maximum number of electrons:

Electron shell	Maximum number of electrons
1st	2
2nd	8
3rd	8

An electron will go into the lowest electron shell which is not fully occupied.

Using a Periodic Table, you should be able to work out the electron shell configurations for the first 20 elements.

The electron shell configuration can be used to identify the Group and Period to which an element belongs.

Magnesium (2.8.2) and calcium (2.8.8.2) have similar chemical properties because they both have two electrons in their outer shell. When they react, they both lose two electrons to become 2+ ions. Calcium is more reactive than magnesium because the outer shell electrons in calcium are further from the nucleus, are less firmly held and so are more easily lost.

2 electrons in the outer shell so Group 2

Mg 2.8.2

3 shells occupied so Period 3

Quick check questions

1 (a) Describe the appearance of an emission spectrum.

   (b) Why do hydrogen atoms give an emission spectrum whereas hydrogen nuclei do not?

2 Draw a diagram of the energy levels in a hydrogen atom. Draw arrows on your diagram to show the origin of three of the lines in the hydrogen absorption spectrum.

3 Explain why hydrogen absorbs only certain frequencies of light.

4 Give the electron structures (in terms of electron shells) for carbon and silicon.

5 Explain why strontium and calcium have similar chemical properties.

6 Explain why potassium is more reactive than sodium.

Chemical bonding and shapes of molecules

Chemical Ideas 3.1 and 3.3

What type of bond?

The type of bond depends on the two atoms involved in the bond.

Between ...	Metal	Non-metal
Metal	metallic bonding	ionic bonding
Non-metal	ionic bonding	covalent bonding

To see where metals and non-metals are situated in the Period Table, see page 3.

Ionic bonding

> Note that you only need to draw the outermost electrons.

Atoms are usually more stable if they have a full outer shell of electrons. The metal atom <u>transfers</u> electron(s) to the non-metal atom so that all atoms have a full outer shell of electrons. This results in the formation of charged ions.

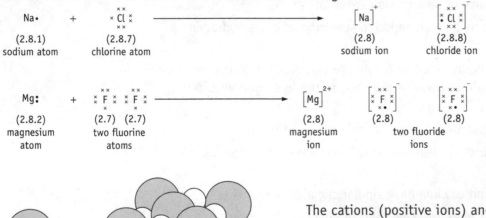

Na•
(2.8.1)
sodium atom

+

× Cl ×
(2.8.7)
chlorine atom

⟶

$[Na]^+$
(2.8)
sodium ion

$\begin{bmatrix} Cl \end{bmatrix}^-$
(2.8.8)
chloride ion

Mg:
(2.8.2)
magnesium atom

+

× F × × F ×
(2.7) (2.7)
two fluorine atoms

⟶

$[Mg]^{2+}$
(2.8)
magnesium ion

$\begin{bmatrix} F \end{bmatrix}^-$
(2.8)

$\begin{bmatrix} F \end{bmatrix}^-$
(2.8)
two fluoride ions

Cl⁻ ion

Na⁺ ion

The cations (positive ions) and anions (negative ions) produced are held together in a giant ionic lattice. There is an electrostatic attraction between the cations and anions.

Covalent bonding

To attain a full outer shell of electrons, the two non-metal atoms involved in a covalent bond <u>share</u> a pair of electrons.

Molecular formula	Dot and cross diagram	Structural formula
H_2	H ⚬ H	H–H
NH_3	H ⚬ N ⚬ with H above and below	H — N: with H above and below
H_2O	H ⚬ O: with H above	H — O: with H above
O_2	O ⚬ O	O = O
N_2	N ⚬ N	N ≡ N
CO_2	O ⚬ C ⚬ O	O = C = O

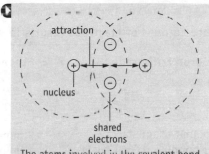

If the two atoms share <u>two</u> pairs of electrons, then a double bond is formed. If they share three pairs of electrons, then a triple bond is formed.

In some molecules, there is a special type of covalent bond called a **dative covalent bond**. Both of the electrons in a dative covalent bond come from the same atom.

dot-cross diagram	an arrow is sometimes used to show a dative bond

Carbon monoxide

Ammonium ion

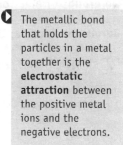

The atoms involved in the covalent bond are held together by an **electrostatic attraction** between the positive nuclei of the two atoms and the shared pair of negative electrons.

Metallic bonding

The metal ions are arranged regularly in a lattice. The outer shell electrons are shared by all the ions and are said to be **delocalised**. The 'sea' of electrons are free to move and because of this, metals conduct electricity.

The metallic bond that holds the particles in a metal together is the **electrostatic attraction** between the positive metal ions and the negative electrons.

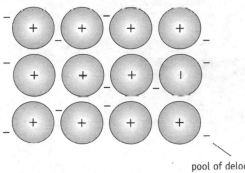

pool of delocalised electrons

Shapes of molecules

- The shape of a molecule depends on the number of **groups** of electrons.
- A group of electrons could be a bonding pair, a non-bonding pair, a double bond (two bonding pairs) or a triple bond (three bonding pairs).
- Groups of electrons **repel** each other.
- Groups of electrons will arrange themselves so **as to be as far apart in space as possible**.

Four groups of electrons results in bond angles of 109°.

Methane, CH_4	Ammonia, NH_3	Ammonium ion, NH_4^+	Water, H_2O	Oxonium ion, H_3O^+
tetrahedral 4 bonding pairs	pyramidal 3 bonding pairs, 1 non-bonding pair	tetrahedral 4 bonding pairs	bent 2 bonding pairs, 2 non-bonding pairs	pyramidal 3 bonding pairs, 1 non-bonding pair
109°		109°		109°

Three groups of electrons results in bond angles of 120°.

Boron trifluoride, BF_3	Ethene, C_2H_4	Methanal, CH_2O
planar triangular 3 bonding pairs	2 bonding pairs, 1 double bond for each C	2 bonding pairs, 1 double bond
F—B 120°	C=C 120°, 120°, 120°	C=O, H each 120°

> Propene's bond angles aren't all the same.
>
> C=C 109°, 120°

Two groups of electrons results in bond angles of 180°.

Beryllium chloride, $BeCl_2$	Ethyne, C_2H_2	Carbon dioxide, CO_2
linear 2 bonding pairs	linear 1 triple bond, 1 bonding pair for each C	linear 2 double bonds
Cl—Be—Cl 180°	H—C≡C—H 180°	O=C=O 180°

? *Quick check questions*

1 Describe the type of bonding in CH_4, KF, SiH_4, Mg and Hg.

2 Draw dot-and-cross diagrams for CH_4, C_2H_6, CH_3CH_2OH, CO_2, CH_2CH_2 and CH_3OCH_3.

3 What are the bond angles **a** to **f** in the following molecules?

Group 2 and balancing equations
Chemical Ideas 1.2 and 11.2

The elements

The Group 2 elements are Be, Mg, Ca, Sr, Ba and Ra.

- The metals react with water to give the metal hydroxide and hydrogen.

metal + water → metal hydroxide + hydrogen

$$M(s) + 2H_2O(l) \rightarrow M(OH)_2(s) \text{ or (aq)} + H_2(g)$$

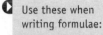

In these equations M is any Group 2 metal.

The oxides

- The oxides react with water to produce an alkaline solution of the hydroxide.

metal oxide + water → metal hydroxide

$$MO(s) + H_2O(l) \rightarrow M(OH)_2(s) \text{ or (aq)}$$

- The oxides react with acids so act as bases.

metal oxide + acid → salt + water

$$MO(s) + H_2SO_4(aq) \rightarrow MSO_4(aq) + H_2O(l)$$

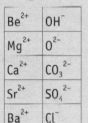

Use these when writing formulae:

Be^{2+}	OH^-
Mg^{2+}	O^{2-}
Ca^{2+}	CO_3^{2-}
Sr^{2+}	SO_4^{2-}
Ba^{2+}	Cl^-

You also need to remember H_2O, CO_2, H_2, HCl and H_2SO_4.

The hydroxides

Become more soluble as you go down the Group, are alkaline in solution since the solutions contain $OH^-(aq)$ (pH > 7) and react with acids to give a salt and water.

metal hydroxide + acid → salt + water

$$M(OH)_2(s) + 2HCl(aq) \rightarrow MCl_2(aq) + H_2O(l)$$

Group 2 elements have similar chemical reactions as they all have two electrons in their outer shell.

The carbonates

Become less soluble as you go down the Group and undergo thermal decomposition on heating to give the metal oxide and carbon dioxide.

metal carbonate → metal oxide + carbon dioxide

$$MCO_3(s) \rightarrow MO(s) + CO_2(g)$$

Thermal stability increases down the Group.

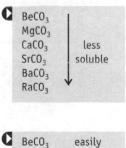

$BeCO_3$
$MgCO_3$
$CaCO_3$ less
$SrCO_3$ soluble
$BaCO_3$
$RaCO_3$

? *Quick check questions*

1 Write an equation for the reaction of strontium with water.

2 Write the formulae for: (a) calcium chloride, (b) strontium hydroxide.

3 Strontium hydroxide solution contains hydroxide ions. How would you test the solution for the presence of hydroxide ions and what would the result be?

4 Write a balanced equation, with state symbols, for the thermal decomposition of calcium carbonate.

5 Which one in each pair is the most soluble? (a) $Mg(OH)_2$ or $Ba(OH)_2$; (b) $MgCO_3$ or $BaCO_3$.

$BeCO_3$ easily decomposed by a Bunsen burner flame

$MgCO_3$
$CaCO_3$ strong heating needed for decomposition

$SrCO_3$
$BaCO_3$
$RaCO_3$

Developing Fuels (DF)

This unit tells the story of petrol: what it is and how it is made. It describes the work of chemists in improving fuels for motor vehicles and in searching for alternative fuels for the future. As you work through this unit you will cover the following concepts. CI refers to sections in your Chemical Ideas textbook.

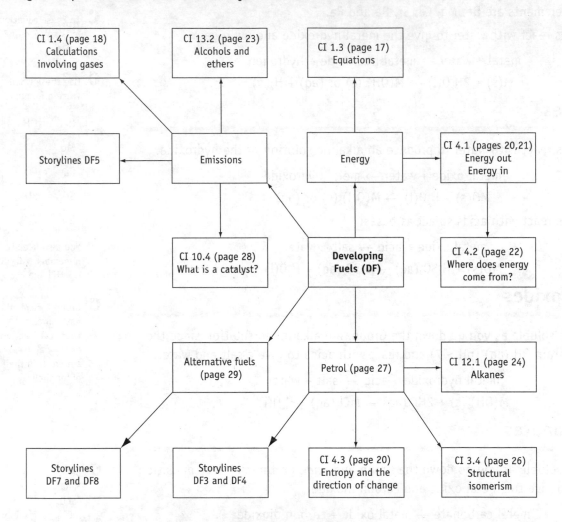

CI 1.4 (page 18)
Calculations involving gases

CI 13.2 (page 23)
Alcohols and ethers

CI 1.3 (page 17)
Equations

Storylines DF5

Emissions

Energy

CI 4.1 (pages 20,21)
Energy out
Energy in

CI 10.4 (page 28)
What is a catalyst?

Developing Fuels (DF)

CI 4.2 (page 22)
Where does energy come from?

Alternative fuels
(page 29)

Petrol (page 27)

CI 12.1 (page 24)
Alkanes

Storylines
DF7 and DF8

Storylines
DF3 and DF4

CI 4.3 (page 20)
Entropy and the direction of change

CI 3.4 (page 26)
Structural isomerism

Calculations from equations
Chemical Ideas 1.3 and 1.4

The large numbers **in front** of formulae in an equation tell you the numbers of moles involved.

$$CH_4 + 2O_2 \rightarrow CO_2 + 2H_2O$$

1 mole of methane reacts with **2 moles** of oxygen to produce **1 mole** of carbon dioxide and **2 moles** of water.

Working out masses

To work out the mass of 1 mole of a compound (M_r), add up the relative atomic mass of each atom in the compound.

e.g. $CaCO_3$ = 40 + 12 + 16 + 16 + 16 = 100

$$moles = \frac{Mass}{Mr}$$

> To work out the mass of 2 moles of a compound, multiply the M_r by 2
>
> mass = amount (moles) × M_r

Worked example

What mass of magnesium oxide is produced when 1.2 g of magnesium reacts with carbon dioxide?

> The relative atomic masses will be given in exam questions.

Step 1 Underline the substances whose: • mass you are given • mass you want to find	2Mg	+ CO_2	→ 2MgO	+ C
Step 2 Indicate moles involved	2 moles		? moles	
Step 3 Calculate the masses	48 g		80 g	
Step 4 Convert to the mass given in the question	$\frac{48}{48} \times 1.2 = 1.2$ g			
Step 5 Convert the other mass by the same amount			$\frac{80}{48} \times 1.2 = 2.0$ g	
Step 6 Write down the answer			2 g of MgO is produced	

Note: **Step 4:** converting the mass

The relative atomic mass of magnesium is 24; there are 2 moles in this equation; 24 × 2 = 48. Dividing by 48 gives 1 g of magnesium; then multiplying by 1.2 gives 1.2 g – as in the question.

Note: **Step 5:** converting the other mass

Converting the other mass by the same amount keeps the proportions correct.

> The approach shown in Steps 4 and 5 has been used in all the worked examples.

Working out volumes

> The volume of one mole of a gaseous substance is 24 dm^2 at room temperature and pressure.

Worked example

What volume of carbon dioxide is produced in the complete combustion of 6.0 dm^3 of methane?

Step 1 Underline the substances whose:
- volume you are given
- volume you want to find

$$\underline{CH_4} \qquad\qquad + 2O_2 \quad \rightarrow \underline{CO_2} \qquad\qquad + 2H_2O$$

> To work out the volume of 2 moles of a substance, multiply 24 by 2.

Step 2 Indicate moles involved

1 mole   1 mole

Step 3 Calculate the volumes   24 dm^3 24 dm^3

Step 4 Convert to the volume given in the question

$\dfrac{24}{24} \times 6 = 6.0$ dm^3

Step 5 Convert the other volume by the same amount

$\dfrac{24}{24} \times 6 = 6.0$ dm^3

> The volume of a mole of gas will be given in exam questions.

Step 6 Write down the answer

6.0 dm^3 of CO_2 is produced

Working with both mass and volume

Worked example

What volume of carbon dioxide is produced when 20 g of calcium carbonate is heated?

Step 1 Underline the substances whose:
- mass or volume you are given
- mass or volume you want to find

$$\underline{CaCO_3} \qquad\qquad \rightarrow \underline{CO_2} \qquad\qquad + CaO$$

Step 2 Indicate moles involved   1 mole   1 mole

Step 3 Calculate the mass or volume   100 g   24 dm^3

Step 4 Convert to the mass or volume given in the question

$\dfrac{100}{100} \times 20 = 20$ g

Step 5 Convert the other mass or volume by the same amount

$\dfrac{24}{100} \times 20 = 4.8$ dm^3

Step 6 Write down the answer

4.8 dm^3 of CO_2 is produced

Working with enthalpy changes

▶ The enthalpy change of combustion tells you the amount of energy released when **1 mole** of a fuel is burnt completely.

Worked example

The enthalpy of combustion for methane is -890 kJ mol^{-1}. Calculate the energy released when 3.2 g of methane burn completely.

Step 1 Underline
- the mass you are given
- the enthalpy change you are given

$$\underline{CH_4} + 2O_2 \rightarrow CO_2 + 2H_2O \quad \Delta H = -\underline{890} \text{ kJ mol}^{-1}$$

▶ To work out energy released when 2 moles of a fuel is burnt, multiply the enthalpy change of combustion by 2.

Step 2 Indicate moles involved — 1 mole — $\Delta H = -890$ kJ mol^{-1}

Step 3 Calculate the mass — 16 g

Step 4 Convert to the mass or volume given in the question
$$\frac{16}{16} \times 3.2 = 3.2 \text{ g}$$

Step 5 Convert the energy released by the same amount
$$\frac{-890}{16} \times 3.2 = -178 \text{ kJ mol}^{-1}$$

Step 6 Write down the answer — 178 kJ of energy is released

▶ Table of relative atomic masses, A_r

H	1
C	12
N	14
O	16
Na	23
Mg	24
Ca	40
Fe	56

Quick check questions

1 What mass of iron(III) oxide is produced when 11.2 g of iron burn?
$$4Fe(s) + 3O_2(g) \rightarrow 2Fe_2O_3(s)$$

2 Calculate the volume of hydrogen which would react exactly with 6 dm^3 of carbon monoxide.
$$6CO(g) + 13H_2(g) \rightarrow C_6H_{14}(l) + 6H_2O(l)$$

3 Calculate the volume of nitrogen gas which would be produced by the thermal decomposition of 0.65 g of sodium azide.
$$2NaN_3(s) \rightarrow 2Na(s) + 3N_2(g)$$

4 The enthalpy change of combustion for octane is -5470 kJ mol^{-1}. How much energy is released when 5.7 g of octane burn completely?
$$C_8H_{18}(l) + 12.5O_2(g) \rightarrow 8CO_2(g) + 9H_2O(l)$$

Enthalpy and entropy
Chemical Ideas 4.1 and 4.3

Enthalpy

- An **exothermic** reaction **gives out energy**. The temperature of the surroundings increases but ΔH is negative because the enthalpy of the reactants decreases.
- An **endothermic** reaction **takes in energy**. The temperature of the surroundings decreases but ΔH is positive because the enthalpy of the reactants increases.
- The **standard enthalpy change of combustion**, $\Delta H_c^{\ominus}$, is the enthalpy change when 1 mole of a substance burns completely in oxygen under standard conditions. $\Delta H_c^{\ominus}$ is always negative.

$$CH_4(g) + 2O_2(g) \rightarrow CO_2(g) + 2H_2O(l); \; \Delta H_c^{\ominus} = -890 \text{ kJ mol}^{-1}$$

- The **standard enthalpy change of formation**, $\Delta H_f^{\ominus}$, is the enthalpy change when 1 mole of a substance is formed from its constituent elements. Both reactants and products are in their standard state. $\Delta H_f^{\ominus}$ can be negative or positive.

> Standard conditions are 1 atmosphere pressure and 298 K (25°C). Note that under these conditions, H_2O is a liquid.

> If the equation is doubled, then so is ΔH.

> Remember to put H_2 or O_2 in equations for enthalpy of formation, not H and O.

Measuring ΔH_c in the laboratory

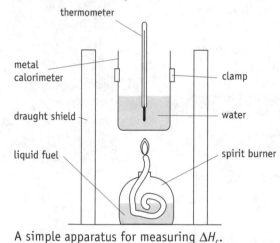

A simple apparatus for measuring ΔH_c.

Record the temperature rise when a known volume of water is heated by the complete combustion of a measured mass of fuel.

m = mass of water (g)

c = specific heat capacity of water $(\text{J g}^{-1} \text{K}^{-1})$

ΔT = change in temperature (K)

The energy transferred = $mc\Delta T$

You can now calculate the enthalpy change for the combustion of 1 mole of fuel.

> Temperature rise is the same in K as in °C. 4.17 J g^{-1} K^{-1} is the specific heat capacity of water.

Entropy

- Entropy is a measure of the number of ways in which particles can be arranged.
- Gases have greater entropy than liquids; liquids have greater entropy than solids.
- Mixtures (e.g. solutions) have greater entropy than the unmixed constituents.
- If the number of particles increases during the course of a reaction then entropy usually increases.

> Don't use the word <u>atom</u> in entropy questions.

Hess's law

Chemical Ideas 4.1

Hess's law

Hess's law states that as long as the starting and finishing points are the same, the enthalpy change will always be the same no matter how you go from start to finish. It is useful for calculating unknown enthalpy changes from ones for which data is available.

Enthalpy cycles

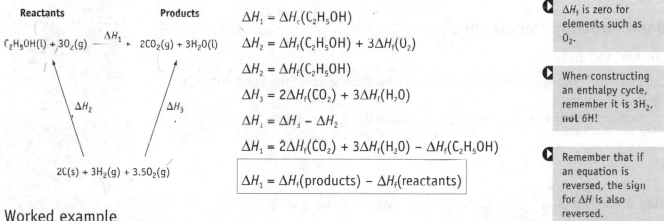

$\Delta H_1 = \Delta H_c(C_2H_5OH)$

$\Delta H_2 = \Delta H_f(C_2H_5OH) + 3\Delta H_f(O_2)$

$\Delta H_2 = \Delta H_f(C_2H_5OH)$

$\Delta H_3 = 2\Delta H_f(CO_2) + 3\Delta H_f(H_2O)$

$\Delta H_1 = \Delta H_3 - \Delta H_2$

$\Delta H_1 = 2\Delta H_f(CO_2) + 3\Delta H_f(H_2O) - \Delta H_f(C_2H_5OH)$

$\boxed{\Delta H_1 = \Delta H_f(\text{products}) - \Delta H_f(\text{reactants})}$

▶ ΔH_f is zero for elements such as O_2.

▶ When constructing an enthalpy cycle, remember it is $3H_2$, not $6H$!

▶ Remember that if an equation is reversed, the sign for ΔH is also reversed.

Worked example

Calculate the standard enthalpy change of combustion for ethanol using the standard enthalpy changes of formation on the right.

$$C_2H_5OH(l) + 3O_2(g) \rightarrow 2CO_2(g) + 3H_2O(l)$$

$\Delta H_c = \Delta H_f(\text{products}) - \Delta H_f(\text{reactants})$

$\Delta H_f(\text{products}) = (2 \times -394) + (3 \times -286) = -1646$

$\Delta H_f(\text{reactants}) = -277$

$\Delta H_c = -1646 - (-277) = -1369 \text{ kJ mol}^{-1}$

The sign and the units may carry a mark. Don't forget them. If the sign is +, then put the +, don't forget it.

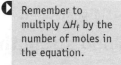

Compound	$\Delta H_f^{\ominus}$, kJ mol^{-1}
$C_2H_5OH(l)$	−277
$CO_2(g)$	−394
$H_2O(l)$	−286

▶ Remember to multiply ΔH_f by the number of moles in the equation.

? Quick check questions

1 Draw and label an enthalpy cycle involving $\Delta H_c(N_2H_4)$, $\Delta H_f(N_2H_4)$ and $\Delta H_f(H_2O)$.

$$N_2H_4(l) + O_2(g) \rightarrow N_2(g) + 2H_2O(l)$$

2 Use the data to calculate $\Delta H_c^{\ominus}$ for carbon monoxide. $\Delta H_f(CO) = -110.5 \text{ kJ mol}^{-1}$
$\Delta H_f(CO_2) = -393.5 \text{ kJ mol}^{-1}$

$$CO(g) + \tfrac{1}{2}O_2(g) \rightarrow CO_2(g)$$

3 Use the data to calculate $\Delta H_c^{\ominus}$ for hydrazine. $\Delta H_f(N_2H_4) = +51 \text{ kJ mol}^{-1}$
$\Delta H_f(H_2O) = -286 \text{ kJ mol}^{-1}$

$$N_2H_4(l) + O_2(g) \rightarrow N_2(g) + 2H_2O(l)$$

Bond enthalpies

Chemical Ideas 4.2

Bond enthalpy

This is the **average** energy required to **break** the bond in 1 mole of **gaseous** compounds.

- Bond breaking is endothermic, ΔH is +ve.
- Bond making is exothermic, ΔH is –ve.

Bond	Bond enthalpy, kJ mol^{-1}
C–C	+347
C–H	+413
O=O	+498
O–H	+464
C=O	+805
C–O	+358
H–H	+436
C=C	+612

Bond enthalpy calculations

Worked example

Calculate the enthalpy change of combustion for ethane using the bond enthalpy data on the right.

$$C_2H_6 \quad + \quad 3.5O_2 \quad \rightarrow \quad 2CO_2 \quad + \quad 3H_2O$$

Step 1 Bonds broken

$1 \times$ C–C (347) = 347
$6 \times$ C–H (413) = 2478
$3.5 \times$ O=O (498) = 1743
Total = **+4568**
(+ sign as bond breaking)

Step 2 Bonds made

$4 \times$ C=O (805) = 3220
$6 \times$ O–H (464) = 2784
Total = **–6004**
(– sign as bond making)

Step 3 $\Delta H_c(C_2H_6) = +4568 - 6004 = -1436$ kJ mol^{-1}
This value is not exactly the same as quoted values, since:
- bond enthalpies are the average energy needed to break that particular bond and is not specific to the molecule in the equation;
- bond enthalpies are for gaseous molecules and this may not be their standard state (the state at 298 K and 1 atmosphere).

It is a good idea to mark the bonds as you count them!

A common mistake is to count two C–C bonds because there are two carbon atoms.

It is a good idea to draw the molecules involved. Learn the structural formulae of O_2, CO_2 and H_2O.

Bond strengths

- The greater the bond enthalpy, the stronger the bond.
- Short bonds are stronger than long bonds.
- C=C is shorter and stronger than C–C.

ΔH_c increases by a regular amount with the addition of each –CH$_2$–. The same additional bonds are broken and made.

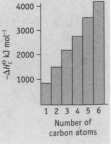

❓ Quick check questions

1 Use bond enthalpy data to calculate a value for the enthalpy of combustion of hydrogen. $H_2 + 0.5O_2 \rightarrow H_2O$

2 Use bond enthalpy data to calculate the enthalpy of combustion of hept-1-ene. $C_7H_{14} + 10.5O_2 \rightarrow 7CO_2 + 7H_2O$

Alcohols and ethers
Chemical Ideas 13.2

Alcohols

- Have an **hydroxyl** (–OH) group and have names ending in **ol**.

Naming alcohols

- Count the number of carbons in the longest chain.
- Replace the **e** at the end of the parent alkane with **ol**.
- Locate the position of the OH group with as **low** a number as possible.
- Remember to retain the **e** for a **diol** or **triol**.

Formulae

Propan–1–ol has a full structured formula of

H–C–C C–O–H

and a skeletal formula of ⌇OH

Combustion

- Alcohols react completely with O_2 to give CO_2 and H_2O.

Oxygenates

- Alcohols require less oxygen for complete combustion than the corresponding alkane since the molecules already have oxygen in them.

Compare $C_2H_5OH + 3O_2 \rightarrow 2CO_2 + 3H_2O$ with $C_2H_6 + 3.5O_2 \rightarrow 2CO_2 + 3H_2O$

- Alcohols (and ethers) are called **oxygenates**; they burn more efficiently than alkanes, producing less carbon monoxide.

Ethers

Have an **alkoxy** group, –OR. e.g. H–C–C–C–O–C–H

This is methoxypropane and **not** propoxymethane.

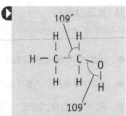

Number of carbons	Alcohol
1	methanol
2	ethanol
3	propanol
4	butanol
5	pentanol
6	hexanol

Correct   Incorrect

The correct skeletal formula will gain 2 marks in an exam question but the incorrect version will only gain 1 mark.

Alkoxy groups

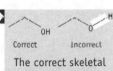

OCH_3	methoxy
OC_2H_5	ethoxy
OC_3H_7	propoxy

? **Quick check questions**

1 Draw the full structural formulae of methanol and ethanol.

2 Name this alcohol:

H–C–C–C–C–C–H (with OH on fourth carbon)

3 Draw the full structural formulae for these alcohols and give their names.

4 Suggest a reason why ethoxyethane is being considered as a possible alternative to petrol in car engines.

Alkanes and other hydrocarbons
Chemical Ideas 12.1, 12.2 and 12.3

Hydrocarbons are compounds of carbon and hydrogen only.

Alkanes

- Have the general formula C_nH_{2n+2}.
- Name ends in **ane**.
- Are **saturated** – all the bonds between carbon atoms are **single bonds**.
- Are **aliphatic** – they don't have a benzene ring.

Naming alkanes

- Choose the longest carbon chain and name it.
- Use prefixes in alphabetical order for any alkyl side chains.
- Use di, tri, tetra before the prefix if the side chains are identical.
- Show the position of any side chains by using numbers which are as low as possible.

Examples

2,2-dimethylheptane   3-ethyl-2-methylheptane

Burning alkanes

- Alkanes react completely with O_2 to produce CO_2 and H_2O.

Different types of formulae

Shortened structural formula	Full structural formula	Skeletal formula
CH₃ │ CH₃—CH—CH₃	(full structural formula of propane)	(skeletal formula)
	If asked for the full structural formula, put bonds between **all** atoms.	You only show the bonds between carbon atoms. Don't put dots for carbons.

Learn these names.

CH_4	**meth**ane
C_2H_6	**eth**ane
C_3H_8	**prop**ane
C_4H_{10}	**but**ane
C_5H_{12}	**pent**ane
C_6H_{14}	**hex**ane
C_7H_{16}	**hept**ane
C_8H_{18}	**oct**ane
C_9H_{20}	**non**ane
$C_{10}H_{22}$	**dec**ane

All the angles in alkanes are the same, 109°. The bonds are arranged tetrahedrally.

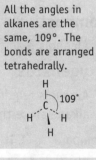

Double check to ensure that you choose the longest chain.

Put a comma between numbers and a hyphen between a number and a letter.

Don't forget to check that each carbon atom in a full structural formula has four bonds coming from it.

Skeletal formulae are most often used for very complicated structures, not simple structures such as ethane and propane.

Other hydrocarbons

Cycloalkanes

- Have the general formula C_nH_{2n}.
- Name ends in **ane**.
- Are **saturated** – all the bonds between carbon atoms are **single bonds**.
- Are **aliphatic** – they don't have a benzene ring.

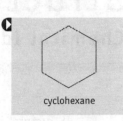

cyclohexane

Alkenes

- Have the general formula C_nH_{2n}.
- Name ends in **ene**.
- Are **unsaturated** – there is a **double bond** between carbon atoms.
- Are **aliphatic** – they don't have a benzene ring.

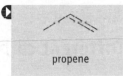

propene

Arenes

- Name ends in **ene**.
- Are **unsaturated**.
- Are **aromatic** – they have a **benzene ring**.

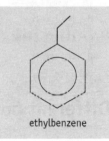

ethylbenzene

? *Quick check questions*

1. Name $CH_3–CH_2–CH_2–CH_2–CH_2–CH_2–CH_2–CH_3$.

2. Draw the skeletal formulae for 3-methylhexane and 2,2,4-trimethylheptane.

3. Write balanced equations for the complete combustion of methane and of ethane.

4. What do the molecular formulae of hexene, 2-methylpentene and cyclohexane have in common?

5. Draw the skeletal formula for the cycloalkane C_5H_{10}. Why is it aliphatic and not aromatic?

Structural isomerism

Chemical Ideas 3.4

> Structural isomers have the same molecular formula but different structural formulae.

Different carbon chains

- This is often seen in alkanes.

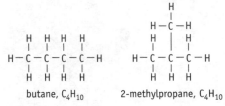

butane, C_4H_{10} 2-methylpropane, C_4H_{10}

Different positions for the functional group

- This is often seen in alcohols.

pentan-1-ol, $C_5H_{12}O$ pentan-2-ol, $C_5H_{12}O$ pentan-3-ol, $C_5H_{12}O$

Different functional groups

- This can be seen in alcohols and ethers.

ethanol, C_2H_6O methoxymethane, C_2H_6O

Isomerisation at the oil refinery

When straight-chain alkanes are heated in the presence of a catalyst, they become branched alkanes. The branched-chain alkanes have a higher octane number and less tendency to auto-ignite.

? *Quick check questions*

1 Give the skeletal formulae of the isomers 2-methylpropan-2-ol and 2-methylpropan-1-ol. Why are they said to be isomers?

2 Give the full structural formulae for two isomers with the molecular formula C_2H_6O. Name the functional group in each of them.

Auto-ignition and octane numbers

Chemical Storylines DF3 and DF4

Definitions

- Auto-ignition is the explosion of a fuel without a spark.
- The octane number is a measure of the tendency of a fuel to auto-ignite.

Octane numbers

100% 2,2,4-trimethylpentane

Octane number 100 $\Longleftrightarrow$

Low tendency to auto-ignite

100% heptane

Octane number 0

High tendency to auto-ignite

If a fuel has an octane number of 80, it has the same tendency to auto-ignite as a mixture of 80% 2,2,4-trimethylpentane and 20% heptane.

Comparing octane numbers

- Short-chain compounds have a higher octane number than long-chain compounds.
- Branched-chain compounds have a higher octane number than long-chain compounds.
- Cycloalkanes have higher octane numbers than straight-chain alkanes.
- Arenes have higher octane numbers than cycloalkanes.
- Oxygenates have higher octane numbers than the corresponding alkanes.

Auto-ignition causes:
- a knocking or 'pinking' sound
- reduces engine performance
- can damage engines.

Remember that a **high** octane number means a **low** tendency to auto-ignite.

Arenes or aromatic compounds have a benzene ring.

benzene

Oxygenates, such as MTBE, have oxygen atoms in the molecule.

MTBE

Cracking, reforming and isomerisation are carried out at the oil refinery to produce organic compounds with higher octane numbers – see pages 26 and 28.

Quick check questions

1 Place the following compounds in order of increasing octane number.

benzene

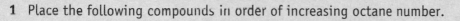

hexane

cyclohexane

2 Which has the lowest octane number?

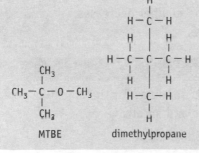

MTBE   dimethylpropane

Catalysts
Chemical Ideas 10.4, Storylines DF4

- A catalyst speeds up a chemical reaction but is not used up in the reaction.
- **Heterogeneous catalysts** are in a different physical state to the reactants.

In a catalytic converter in a car, platinum or rhodium act as heterogeneous catalysts for the following reaction:

$$2NO(g) + 2CO(g) \rightarrow N_2(g) + 2CO_2(g)$$

1 Reactants are **adsorbed** onto the surface of the catalyst.

2 Bonds between atoms inside the reactants **weaken** and are **broken**.

3 New bonds and compounds are **formed**.

4 The products **diffuse** away.

> ▶ If states are required in the answer to a question, you will be told in the question.

> ▶ Catalysts in catalytic converters only work when the temperature is high. They don't work at the start of a journey.

Catalysts and petrol components

Catalysts are used in three processes at the oil refinery to produce petrol components with a **low tendency** to auto-ignite. The products of these three processes have a **higher octane number** than the reactants in the processes.

> ▶ If a substance is adsorbed onto the catalyst surface, it bonds to the catalyst surface.

Cracking

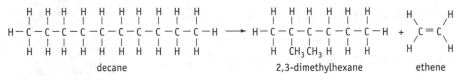

decane   2,3-dimethylhexane   ethene

> ▶ A catalyst poison adsorbs onto the catalyst surface and stops it working.

- A shorter alkane and an alkene are usually formed.

> ▶ The alkane produced is short, so has a higher octane number.

Reforming

- Alkanes are converted to cycloalkanes and hydrogen.
- Cycloalkanes are converted to arenes and hydrogen.
- In terms of octane numbers, arenes > cycloalkanes > alkanes.

Isomerisation

$$CH_3-CH_2-CH_2-CH_2-CH_3 \xrightarrow[\text{zeolite sieve}]{\text{Pt catalyst}} CH_3-\overset{\overset{\displaystyle CH_3}{\displaystyle |}}{CH}-CH_2-CH_3$$

pentane, octane number 62   2-methylbutane, octane number 93

> ▶ The product is shorter and more branched than the reactant. Both factors increase octane numbers.

- The longest carbon chain becomes shorter so the octane number decreases.
- There is more branching so the octane number increases.

❓ Quick check questions

1 Write a balanced chemical equation (using molecular formula) for the cracking of octane to give an alkene with three carbon atoms and an alkane.

2 $C_7H_{16} \rightarrow C_7H_{14} + H_2$ Is this cracking, reforming or isomerisation?

Pollution from a car
Chemical Storylines DF5, DF7 and DF8

Production and effects of pollutants

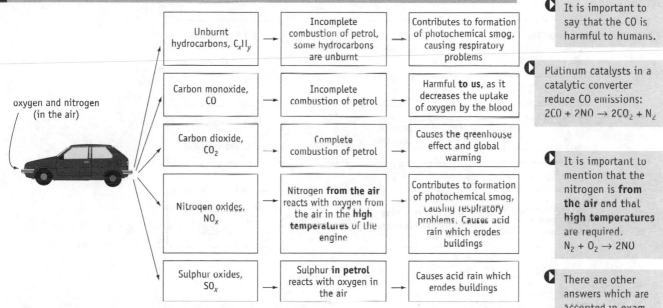

▶ It is important to say that the CO is harmful to humans.

▶ Platinum catalysts in a catalytic converter reduce CO emissions:
$2CO + 2NO \rightarrow 2CO_2 + N_2$

▶ It is important to mention that the nitrogen is **from the air** and that **high temperatures** are required.
$N_2 + O_2 \rightarrow 2NO$

▶ There are other answers which are accepted in exam mark schemes but to avoid confusion, learn these.

▶ Energy density (in $kJ\ kg^{-1}$) – enthalpy of combustion (in $kJ\ mol^{-1}$) × amount of fuel in 1 kg (in $mol\ kg^{-1}$).

▶ Alcohols and ethers are oxygenates. Since they have oxygen atoms in them, their combustion is more complete than the combustion of the corresponding alkane. Less CO is produced.

The ideal fuel for a car engine

- Should have a high energy density in $kJ\ kg^{-1}$.
- Should have a high octane number so that there is a low tendency to auto-ignite.
- Should produce as few pollutants as possible.
- Should have a suitable boiling point. The fuel should be liquid in the tank, should vaporise to become a gas for combustion in the engine but shouldn't evaporate too easily or the fuel will be lost.

Hydrogen has great potential as a fuel since it has a high energy density and only produces water on burning. It must be liquefied however for the fuel tank, which would require either refrigeration or high pressures.

Ethanol has been produced from fermentation of sugar cane and is used as a fuel in Brazil. As well as being a renewable fuel, it has a high octane number and is an oxygenate so produces less carbon monoxide than the hydrocarbons in petrol.

Quick check questions

1 Explain how carbon monoxide is produced.

2 Describe one polluting effect of nitrogen monoxide.

3 A 'lean burn' engine uses a higher ratio of air to petrol vapour than other engines. Why are less unburnt hydrocarbons produced?

4 What problems may arise in the storage of hydrogen in a fuel tank?

5 243 kJ of energy is produced if 1 mole of hydrogen is burnt. What is the energy density of hydrogen in $kJ\ kg^{-1}$.

From Minerals to Elements (M)

This unit looks at the extraction of two very different elements. The first is the non-metal bromine and the second is the metal copper. The following concepts are developed as you follow the storyline. CI refers to sections in your Chemical Ideas textbook.

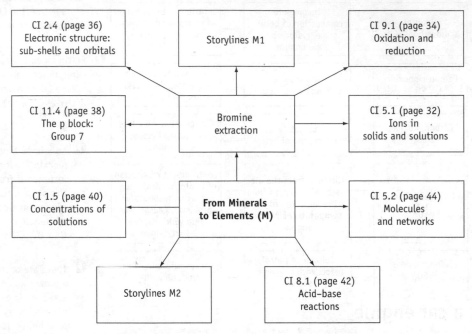

From Chemical Storylines, you also need to be aware of the following.

The extraction of bromine from sea water

Sea water containing bromide ions is filtered, acidified and injected with chlorine. This oxidises the bromide ions to bromine.

1 Oxidation of bromide ions (Br^-) to bromine (Br_2)

$$2Br^-(aq) \rightarrow Br_2(aq) + 2e^-$$ this is oxidation (electron loss)

$$Cl_2(aq) + 2e^- \rightarrow 2Cl^- aq)$$ this is reduction (electron gain)

2 Removal of bromine vapour

Air is blown up a 'blowing-out tower' to remove the volatile bromine from the water.

3 Reduction of bromine to hydrobromic acid (HBr)

Sulphur dioxide is injected into the tower to reduce bromine to hydrogen bromide and also produce sulphuric acid.

$$Br_2(aq) + SO_2(g) + 2H_2O(l) \rightarrow 2HBr(aq) + H_2SO_4(aq)$$

4 Oxidation of hydrobromic acid to bromine

In the 'steaming-out tower' chlorine regenerates the bromine by oxidising the bromide ions from hydrogen bromide.

$$2HBr(aq) + Cl_2(g) \rightarrow Br_2(g) + 2HCl(aq)$$

◗ You do not need to learn the equations in this section but you do need to be able to interpret and use them in an examination question.

◗ Steps 3 and 4 increase the concentration of the bromine produced.

◗ $Br^-(aq)$ is the reducing agent (donates electrons); $Cl_2(aq)$ is the oxidising agent (receives electrons).

Extraction of metals

Chemical Storylines M2

Copper is needed in large amounts for the electrical industry and for piping. The four stages in the extraction of copper are mining, concentration, smelting and electrolytic refining.

Mining

Chalcopyrite ($CuFeS_2$) is the copper mineral mined on a huge scale in Salt Lake City.

Concentration

The rock is ground to a powder. This undergoes froth flotation.

A chemical called a collector is added which binds to the surface of the mineral grains. This makes them water-repelling. Air and detergent are blown in, causing froth. Because they are water repelling, the mineral grains become concentrated in the froth and can be removed. The grains go to the smelter, the waste slurry to the tailings pond.

Smelting

- In flash smelting, the concentrate, silica and oxygen-enriched air is passed down a reaction shaft. A violent exothermic reaction takes place.
- The copper is converted to copper(I) sulphide, known as copper matte, Cu_2S. This is 'blown' with air to produce copper.

$$Cu_2S(l) + O_2(g) \rightarrow 2Cu(l) + SO_2(g)$$

- The sulphur dioxide produced is collected and converted to sulphuric acid. This reduces the amount of acid rain produced.

Electrolytic refining

- The electrolyte is a warm solution of copper(II) sulphate and sulphuric acid.
- The anodes are the impure copper and the cathodes are thin sheets of pure copper.

  The half-equations for the reactions are:

$$Cu^{2+}(aq) + 2e^- \rightarrow Cu(s) \qquad \text{reduction (at the cathode)}$$
$$Cu(s) \rightarrow Cu^{2+}(aq) + 2e^- \qquad \text{oxidation (at the anode)}$$

In large-scale metal extraction there are always environmental factors to consider. Examples are noise pollution, reduction of effluent gases and visual impact of spoil heaps, as well as the treatment of effluent gases. All of these need to be controlled in order to minimise environmental damage.

Ions in solids and solutions

Chemical Ideas 5.1

Structure of an ionic lattice: sodium chloride

Sodium chloride consists of sodium ions (Na^+), each of which is surrounded by six chloride ions (Cl^-). In turn, each Cl^- ion is surrounded by six Na^+ ions. The ions are held together by attractions between these oppositely charged ions. This allows a **giant ionic lattice** to be built up. The lattice structure of sodium chloride is said to be **simple cubic**. Similar electrostatic attractions hold all ionic lattices together.

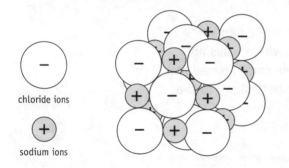

chloride ions

sodium ions

> You may be asked to draw the structure of an ionic compound. Remember to label the ions and their charges.

Sometimes ionic crystals contain some water molecules. These water molecules sit within the ionic lattice and are known as **water of crystallisation**. The crystals are said to be **hydrated**. You may have seen blue copper sulphate crystals. These are in fact the pentahydrate crystal, $CuSO_4.5H_2O$. Anhydrous copper sulphate ($CuSO_4$) is white.

Precipitation reactions

If two solutions react to form a solid, a precipitation reaction is said to have occurred. For example:

$$Ag^+(aq) + NO_3^-(aq) + Na^+(aq) + Br^-(aq) \rightarrow AgBr(s) + Na^+(aq) + NO_3^-(aq)$$

$Na^+(aq)$ and $NO_3^-(aq)$ are spectator ions (i.e. they are not involved in the reaction), so an ionic equation can be written as:

$$Ag^+(aq) + Br^-(aq) \rightarrow AgBr(s)$$

All silver halides are insoluble. For their characteristic colours see page 38.

> Always put in the state symbols for this type of reaction.

Hydration of ions in aqueous solution

Many ionic compounds dissolve in water without difficulty. Whether or not dissolving happens depends on the energy changes involved. The stages involved in dissolving are:

1 ion separation

2 ion hydration (surrounding the ions by water molecules).

The electrostatic attractions in an ionic lattice are large so why do ionic lattices break down and dissolve?

The electrostatic attractions between oppositely charged ions are overcome by the many small charges on the water molecules. Each water molecule is made up of H and O atoms, which have different electronegativities. Water also has a 'bent' shape. This means the water molecule has a small dipole.

Polarity in a water molecule.

> See page 61 for more about electronegativity.

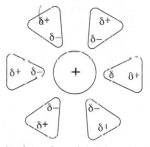

Hydrated positive ion.

The negative end of a water dipole (i.e. the oxygen end) is attracted to a positive ion in the lattice and 'pulls' it away from the lattice. The positive ion is then surrounded with other water molecules, with the oxygen end of each water molecule near to the positive ion.

> If you are asked to draw a labelled diagram of a hydrated ion always label the charge of the ion, show the polarity of the water molecules and surround the ion with a minimum of five water molecules.

The positive end of a water dipole (i.e. the hydrogen end) is attracted to a negative ion in the lattice and 'pulls' it away from the lattice. The negative ion is then surrounded with other water molecules, with the hydrogen end of each water molecule next to the negative ion.

The ions are said to be **hydrated**.

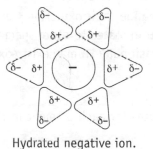

Hydrated negative ion.

? Quick check questions

1 What structure is the lattice of sodium chloride?

2 Write out an ionic equation for the reaction between barium chloride and potassium sulphate (barium sulphate is insoluble in water).

3 Draw a diagram of a potassium ion hydrated by five water molecules, clearly showing the direction of the dipole in water.

Oxidation and reduction

Chemical Ideas 9.1 and Chemical Storylines M1

When an oxidation reaction and a reduction reaction occur simultaneously this is known as a **redox reaction**.

Oxidation and reduction can be defined in two different ways:

- **O**xidation **i**s the **l**oss of electrons.
- **R**eduction **i**s the **g**ain of electrons.

or

- An element is oxidised when its oxidation state is increased (becomes more positive).
- An element is reduced when its oxidation state is decreased (becomes more negative).

> **Remember oil rig:**
> **o**xidation
> **i**s
> **l**oss
> **r**eduction
> **i**s
> **g**ain

Assigning oxidation states

- The atoms in elements are always in an oxidation state of zero.
- In compounds or ions, oxidation states (or numbers) are assigned to each atom or ion. Since compounds have no overall charge, the oxidation states of all the constituents must add up to zero.
- Some atoms rarely change their oxidation states in reactions. These can be used to help assign oxidation states to other species. Examples are: F is −1, O is −2, H is +1, Cl is −1. Occasionally H⁻ may exist (oxidation state −1).

Worked examples

Assign the oxidation states for each element in the following compounds.

(a) CO_2
Step 1 O is −2; there are two so the total contribution of O to the oxidation state is $2 \times (-2) = -4$.

Step 2 To make the total of the oxidation states add up to zero, C must be +4.

(b) CH_4
Step 1 H is +1; there are four so the total contribution of H to the oxidation state is $4 \times (+1) = +4$.
Step 2 To make the total of the oxidation states add up to zero, C is −4.

(c) VO^{2+}
Step 1 O is −2.
Step 2 The charge on the ion is +2, so V must have a charge of +4 as $+4 + (-2) = +2$.

> You must always write down the sign of an oxidation state (+ or −) or you will lose marks.

Using oxidation states

$$Cl_2(aq) + 2I^-(aq) \rightarrow 2Cl^-(aq) + I_2(aq)$$

oxidation states: 0   −1   −1   0

- The oxidation state for chlorine decreases (from 0 to −1), so chlorine is reduced.
- The oxidation state for iodine increases (from −1 to 0), so iodine is oxidised.

Electron transfer and half-equations

Sodium reacts with chlorine as follows:

$$2Na + Cl_2 \rightarrow 2NaCl$$

This can be written as two separate half-equations. From these you can decide which is the oxidation reaction and which is the reduction reaction:

$2Na \rightarrow 2Na^+ + 2e^-$ this is oxidation (electron loss)

$Cl_2 + 2e^- \rightarrow 2Cl^-$ this is reduction (electron gain)

You can also look at how the oxidation number changes:

- Na changes from 0 to +1 and has therefore been oxidised;
- Cl_2 changes from 0 to −1 and has therefore been reduced.

Other redox changes with halide ions

A displacement reaction occurs when a more reactive halogen is passed into a solution of less reactive halide ions.

Example: $Cl_2(g) + 2KI(aq) \rightarrow 2KCl(aq) + I_2(aq)$

$$Cl_2(g) + 2I^-(aq) \rightarrow 2Cl^-(aq) + I_2(aq)$$

> K⁺ is a spectator ion, so it can be removed from the equation.

This reaction can be represented by two half-equations:

$Cl_2(aq) + 2e^- \rightarrow 2Cl^-(aq)$ this is reduction (electron gain)

$2I^-(aq) \rightarrow I_2(aq) + 2e^-$ this is oxidation (electron loss)

> The chlorine molecules are acting as the oxidising agent (they allow another species to be oxidised and in doing so are reduced themselves). The iodide ions are reducing agents (they allow another species to be reduced and in doing so are oxidised themselves).

Quick check questions

1 Write down the oxidation states of the elements in the following:
KBr, H_2O, PO_4^{3-}

2 Identify the changes in oxidation states in the following reaction:
$$2Br^- + 2H^+ + H_2SO_4 \rightarrow Br_2 + SO_2 + 2H_2O$$

3 (a) Write two half-equations for the following reaction:
$$2Ca + O_2 \rightarrow 2CaO$$
(b) Identify which equation is the oxidation reaction and which is the reduction reaction.

Electronic structure: sub-shells and orbitals

Chemical Ideas 2.4

- Electrons exist in shells and these are designated $n = 1$, $n = 2$, $n = 3$, etc. The further a shell is away from the nucleus the larger its n number.
- These shells are sub-divided into sub-shells designated s, p, d and f.
- Each sub-shell is further divided into atomic orbitals. Each atomic orbital can hold a maximum of two electrons. These two electrons are oppositely spin paired.

'Arrows in boxes' are a good way to represent the filling up of orbitals by electrons.

Distribution of electrons in atomic orbitals

The first shell	$n = 1$ has only one sub-shell	s
The second shell	$n = 2$ has two sub-shells	s and p
The third shell	$n = 3$ has three sub-shells	s, p and d
The fourth shell	$n = 4$ has four sub-shells	s, p, d and f

- An s sub-shell has 1 orbital (a maximum of 2 electrons)
- A p sub-shell has 3 orbitals (a maximum of 6 electrons)
- A d sub-shell has 5 orbitals (a maximum of 10 electrons)

The arrangement of electrons in shells and orbitals is called the **electronic configuration**.

Rules for filling up atomic orbitals

1 The orbitals are filled in order of increasing energy.

2 Where there is more than one orbital at the same energy, the orbitals are first occupied singly by electrons. When each orbital is singly occupied electrons then pair up in orbitals.

3 Electrons in singly occupied orbitals have parallel spins.

4 Electrons in doubly occupied orbitals have opposite spins.

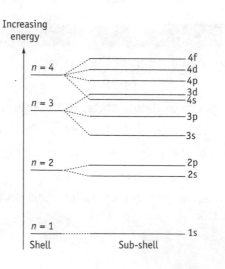

Note the 4s orbital is at a lower energy level than the 3d and so fills up before the 3d.

Representing electron distribution

Each orbital can be represented as a box and each electron as an arrow. The electronic configurations for nitrogen and sodium can be represented as follows:

N 1s $\uparrow\downarrow$ 2s $\uparrow\downarrow$ 2p $\uparrow$ $\uparrow$ $\uparrow$ or $1s^2 2s^2 2p^3$

Na 1s $\uparrow\downarrow$ 2s $\uparrow\downarrow$ 2p $\uparrow\downarrow\uparrow\downarrow\uparrow\downarrow$ 3s $\uparrow$ or $1s^2 2s^2 2p^6 3s^1$

You may be asked to draw diagrams for any of the elements up to and including krypton (atomic number 36). It would be a good idea to practice these.

Deducing electronic configurations

Using the energy level diagram on the previous page, the electron configuration of an element can be deduced.

Worked example

Step 1 Magnesium has an atomic number of 12, and therefore has 12 protons.

Step 2 In atom form magnesium has 12 electrons.
It has 2 in shell $n = 1$, 8 in shell $n = 2$ and 2 in shell $n = 3$.

Step 3 These are further divided into sub-shells written as $1s^2 \ 2s^2 \ 2p^6 \ 3s^2$.

The notation for phosphorus, atomic number 15, is 2, 8, 5 in shells and $1s^2 \ 2s^2 \ 2p^6 \ 3s^2 \ 3p^3$ in sub-shells.
Potassium is $1s^2 \ 2s^2 \ 2p^6 \ 3s^2 \ 3p^6 \ 4s^1$.

s, p and d blocks

- Group 1 and 2 elements all have one or two electrons, respectively, in their outermost sub-shell, which is an s orbital. These are known as the s block elements.

- Group 3, 4, 5, 6, 7 and 0 elements all have three, four, five, six, seven or eight electrons, respectively, in their outermost sub-shell, which are p orbitals. They are known as the p block elements.

- The first transition block elements have electrons which are filling the d sub-shell. These are known as the d block elements.

- Further transition metals have electrons which are filling the f sub-shell. These are known as the f block elements.

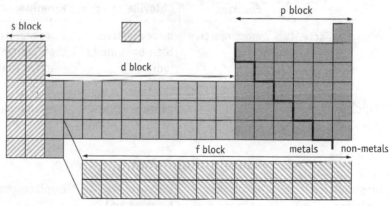

? **Quick check questions**

1 (a) How many sub-shells does the third shell have?
  (b) Name them.

2 What is the electronic configuration for sulphur, atomic number 16?

3 Which element has the electronic configuration $1s^2 \ 2s^2 \ 2p^6 \ 3s^2 \ 3p^6 \ 4s^2$?

There is a Periodic Table on page 3.

The p block: Group 7
Chemical Ideas 11.4

Physical properties

The elements in Group 7 are called the halogens.

You need to be able to recall the following physical properties of the halogens:

	Fluorine	Chlorine	Bromine	Iodine
Appearance and state at room temperature	pale yellow gas	green gas	dark red liquid	shiny black solid
Volatility	pale yellow gas	green gas	brown gas on warming	sublimes on warming to give a purple vapour
Solubility in water	reacts with water	slightly soluble to give pale green solution	slightly more soluble to give red-brown solution	barely soluble, gives a brown solution
Solubility in organic solvents	soluble	soluble to give a pale green solution	soluble to give a red solution	soluble to give a violet solution

You also need to know and be able to explain the following chemical properties of the halogens:

	Fluorine	Chlorine	Bromine	Iodine
Relative reactivity	most reactive	more reactive than bromine or iodine	more reactive than iodine less reactive than chlorine	least reactive
Halide ions and silver ions		white precipitate of silver chloride	cream precipitate of silver bromide	yellow precipitate of silver iodide
	$Ag^+(aq) + X^-(aq) \rightarrow AgX(s)$			
Displacement reactions	displaces chlorine, bromine and iodine	displaces bromine and iodine	displaces iodine	does not displace bromine or chlorine
	$Cl_2(aq) + 2Br^-(aq) \rightarrow 2Cl^-(aq) + Br_2(aq)$ $Cl_2(aq) + 2I^-(aq) \rightarrow 2Cl^-(aq) + I_2(aq)$ $Br_2(aq) + 2I^-(aq) \rightarrow 2Br^-(aq) + I_2(aq)$			
Redox reactions	Halogens are all reactive. They tend to remove electrons from other elements. They are oxidising agents. $X_2 + 2e^- \rightarrow 2X^-$ Halogen oxidation state 0 reduced to oxidation state −1.			

Health and safety precautions

Bromine is transported in lead-lined steel tanks supported in strong metal frames. There are international regulations that control the design and construction of road and rail tankers in order to reduce accidents and the danger to public health. Transport routes are planned to minimise the risk of accidents: e.g. routes are planned to avoid residential areas and often the liquid is transported at night to avoid other traffic.

Economic importance

- Bromine is used in the manufacture of flame retardants, agricultural fumigants, anti-knock agents in leaded petrol and in photography.
- Chlorine (manufactured by the electrolysis of brine) is an important intermediate in the manufacture of hydrochloric acid and chlorinated solvents and is used in the plastics industry.

? Quick check questions

1. What is the appearance of bromine at room temperature?
2. Write the equation for the reaction of potassium iodide solution with silver nitrate. What would you see?
3. (a) Write an ionic equation for the reaction that occurs when chlorine gas is bubbled through potassium bromide solution.

   (b) Describe and explain what you would see.

   (c) If cyclohexane (an organic solvent) was added after the reaction in part (a), what would you see? Explain your answer.

Concentrations of solutions

Chemical Ideas 1.5

Calculations of concentrations

Concentrations can be measured in grams per cubic decimetre (g dm^{-3}).

Example A solution containing 40 g of sodium hydroxide in 2 dm^3 of solution has a concentration of 40/2 = 20 g dm^{-3}.

However, chemists measure concentrations by the 'amount of substance', known as the mole.

1 mole of a substance dissolved in 1 dm^3 of solution has a concentration of 1 mol dm^{-3}.

Example A memory aid can be used to help with concentration calculations.

- concentration (*c*) in mol dm^{-3}
- amount of substance (*n*) in moles
- volume of solution (*V*) in dm^3

$$c = \frac{n}{V} \qquad n = c \times V \qquad V = \frac{n}{c}$$

Any of the three quantities (concentration, amount or volume) can be calculated by using the correct expression and the other two known values.

Examples

Concentration	Amount	Volume
$c = \dfrac{n}{V}$	$n = c \times V$	$V = \dfrac{n}{c}$
0.5 mole of NaOH in 100 cm^3 (100 cm^3 is 0.1 dm^3) $c = \dfrac{0.5}{0.1}$ = 5 mol dm^{-3}	How many moles in 20 cm^3 of 0.1 mol dm^{-3} NaOH? (20 cm^3 is 0.02 dm^3) $n = 0.1 \times 0.02$ = 0.002 mol	What volume of 0.2 mol dm^{-3} NaOH contains 0.1 mol? $V = \dfrac{0.1}{0.2}$ = 0.5 dm^3
unit: mol dm^{-3}	unit: mol	unit: dm^3

> ▶ 1 dm^3 = 1000 cm^3

> ▶ 1 mole of a substance is equal to its relative formula mass in grams.

> ▶ Always remember to convert volumes to dm^3 and quote the correct units for your answer.

> ▶ Do not forget the units.

Using concentrations in calculations

A titration is a method of quantitatively finding the concentration of a solution by reacting a known volume of it with another solution of known concentration. The end-point of the reaction is often seen by the use of an indicator that changes colour.

- A fixed volume of solution of unknown concentration is placed in a conical flask using a pipette.

- A few drops of suitable indicator are added and the mixture placed on a white tile (in order to see the end-point clearly).

- The solution of known concentration is added slowly from a burette, with constant swirling.

- After a rough titration accurate titrations follow until concordant results are obtained.

Worked example

In a titration, 25 cm^3 of potassium hydroxide solution was pipetted into a conical flask. A 0.02 mol dm^{-3} solution of sulphuric acid was added from a burette. An indicator in the flask changed colour when 27.9 cm^3 of sulphuric acid had been added. What is the concentration of the potassium hydroxide?

Step 1 Write down the equation:

$$H_2SO_4 + 2KOH \rightarrow K_2SO_4 + 2H_2O$$

Step 2 Find the ratio of the reactants:
1 mole of sulphuric acid reacts with 2 moles of potassium hydroxide.

Step 3 Using the equation $n = c \times V$, you can calculate the number of moles of H_2SO_4:

$$n(H_2SO_4) = 0.02 \times \frac{27.9}{1000} = 5.58 \times 10^{-4} \text{ moles}$$

> Volume of H_2SO_4
> = 27.9 cm^3
> = $\frac{27.9}{1000}$ dm^3

Step 4 Use the ratio from step 2:
This is equivalent to $2 \times 5.58 \times 10^{-4} = 1.17 \times 10^{-3}$ moles of KOH.

> 1 mole of H_2SO_4
> reacts with 2 moles
> of KOH.

Step 5 Calculate the concentration:
This number of moles is in a volume of 25 cm^3. You can calculate the concentration of the KOH solution from the equation $c = \frac{n}{V} = \frac{1.17 \times 10^{-3}}{0.025} = 0.046$ mol dm^{-3}

> Volume of KOH
> = 25.0 cm^3
> = $\frac{25.0}{1000}$ dm^3
> = 0.025 dm^3

? Quick check questions

1 A volume of 20 cm^3 sulphuric acid was used from a burette. The concentration of the solution was 0.100 mol dm^{-3}.

(a) What was the volume of acid used, in dm^3?

(b) What was the number of moles of acid used?

(c) How many moles of NaOH would this acid neutralise?

$$H_2SO_4(aq) + 2NaOH(aq) \rightarrow Na_2SO_4(aq) + 2H_2O(l)$$

(d) If this number of moles of NaOH were in a 25 cm^3 portion drawn by a pipette and placed in a conical flask, how many moles would there have been in 1000 cm^3 of the solution?

(e) What is the concentration of the sodium hydroxide solution, in mol dm^{-3}?

Acid–base reactions

Chemical Ideas 8.1

One way of characterising acids is their ability to transfer H^+ ions (H^+ is a proton). A base is a substance that accepts H^+ ions. This is the Brønsted–Lowry theory.

dative bond, formed by two electrons

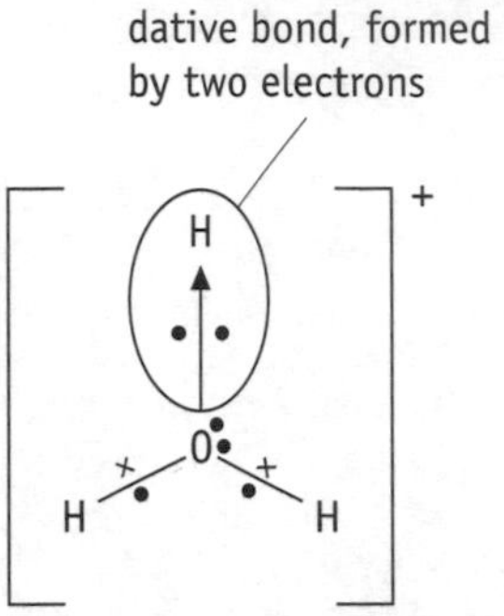

Acids donate H^+ to water in aqueous solution to become H_3O^+ (the **oxonium** ion).

An alkali is a base that dissolves in water to produce hydroxide ions, e.g. sodium hydroxide, potassium hydroxide and sodium carbonate.

Sodium carbonate produces hydroxide ions by the following reaction:

$$CO_3^{2-}(aq) + H_2O(l) \rightleftharpoons HCO_3^-(aq) + OH^-(aq)$$

> ▶ Acidic solutions:
> turn litmus red,
> are neutralised by bases,
> have a pH < 7,
> liberate CO_2 from carbonates.

Proton transfer

In the reaction below a proton is transferred from HNO_3 to H_2O. HNO_3 is acting as the acid and H_2O is acting as the base.

$$H)NO_3 + H_2O \longrightarrow NO_3^- + H_3O^+$$
acid   base

> ▶ An acid is a proton donor and a base is a proton acceptor.

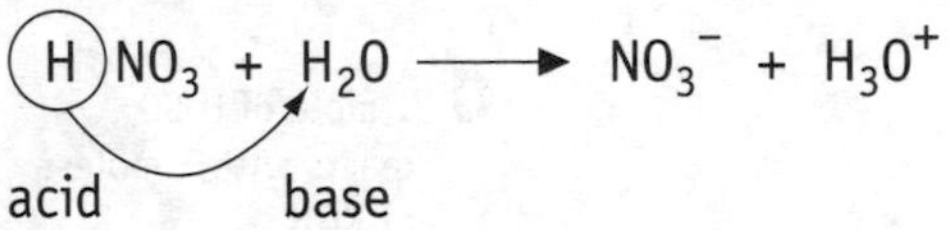

Acid–base pairs

In many cases the donation of a proton by an acid is reversible.

$$HA \rightleftharpoons H^+ + A^-$$
conjugate acid   conjugate base

In the equation above HA donates protons and acts as an acid. A^- acts as a base in the reverse reaction. They are called a conjugate acid–base pair.

Identifying reactions

You are expected to be able to identify the following different types of reaction:

Acid–base

A proton donor and proton acceptor are identified.

Example

$$NH_4^+ + CO_3^{2-} \rightarrow NH_3 + HCO_3^-$$
proton donor   proton acceptor

Redox

Look for changes in oxidation states. One substance loses electrons and is oxidised. Another substance gains electrons and is reduced.

Example

$$H_2 + Cl_2 \rightarrow 2HCl$$

Oxidation state $\qquad$ 0 $\quad$ 0 $\quad\quad$ +1 –1

H has been oxidised from 0 to +1 and Cl has been reduced from 0 to –1. For details see the section on Chemical Ideas 9.1 (pages 34 and 35).

Precipitation

When two solutions react to form a solid this is a precipitation reaction. Look at the state symbols in the equation. The reactants will all be (aq) and one of the products will be (s).

Example

$$Ag^+(aq) + NO_3^-(aq) + Na^+(aq) + Cl^-(aq) \rightarrow AgCl(s) + Na^+(aq) + NO_3^-(aq)$$

$Na^+(aq)$ and $NO_3^-(aq)$ are spectator ions, so an ionic equation can be written as:

$$Ag^+(aq) + Cl^-(aq) \rightarrow AgCl(s)$$

Quick check questions

1 Identify the acid and the base in the following reactions:

(a) $NH_3 + HBr \rightarrow NH_4^+ + Br^-$

(b) $SO_4^{2-} + H_3O^+ \rightarrow HSO_4^- + H_2O$

2 In the reaction below identify the two conjugate acid–base pairs:

$$H_2SO_4 + OH^- \rightarrow HSO_4^- + H_2O$$

3 Classify the following reactions as acid–base, redox or precipitation:

(a) $Ba^{2+}(aq) + SO_4^{2-}(aq) \rightarrow BaSO_4(s)$

(b) $CH_3COOH + H_2O \rightarrow CH_3COO^- + H_3O^+$

(c) $2FeCl_2 + Cl_2 \rightarrow 2FeCl_3$

Molecules and networks
Chemical Ideas 5.2

Some elements and compounds form giant structures, covalently bonded together. We call these **network** or **giant structures**. Two examples are diamond and silicon(IV) oxide.

Diamond

- Diamond is made of carbon atoms.
- Each carbon atom is joined tetrahedrally to four other carbon atoms, by strong covalent bonds.
- The very strong C–C bonds in four different directions make diamond the hardest naturally occurring substance.

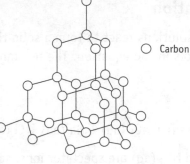

○ Carbon

The structure of diamond.

Silicon(IV) oxide

- Silicon atoms form four bonds.
- Silicon bonds covalently to four oxygen atoms.
- Each silicon atom has a half-share in four oxygen atoms.

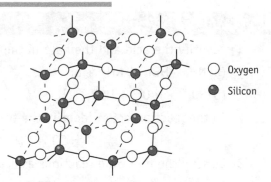

○ Oxygen
● Silicon

The structure of silicon(IV) oxide (SiO_2).

Differences between CO_2 and SiO_2

Carbon atoms are smaller than silicon atoms and so double bonds can form between two carbon atoms. This allows the formation of discrete molecules. Carbon dioxide has a small molecular structure with bonding O=C=O. There are weak intermolecular forces (the forces *between* molecules). The molecules are easily pulled apart, so carbon dioxide has a low melting point and a low boiling point. The bonds in carbon dioxide are polar so it dissolves easily in water.

Polarity of carbon dioxide.

$$\overset{\delta-}{O} = \overset{\delta+}{C} = \overset{\delta-}{O}$$

On the other hand, silicon(IV) oxide has a giant network structure. It requires a lot of energy to pull these intramolecular bonds apart (forces *within* the molecules). Consequently, silicon(IV) oxide has a high melting point and a high boiling point and does not dissolve easily in water.

? Quick check questions

1 Why is diamond the hardest naturally occurring substance?
2 Why is CO_2 a gas at room temperature whilst SiO_2 is a solid?

The Atmosphere (A)

This unit looks at the chemical and physical processes that go on in the atmosphere. It concentrates firstly at looking at the chemistry that is responsible for the depletion of the ozone layer and secondly on global warming. The concepts covered are given below. CI refers to sections in your Chemical Ideas textbook.

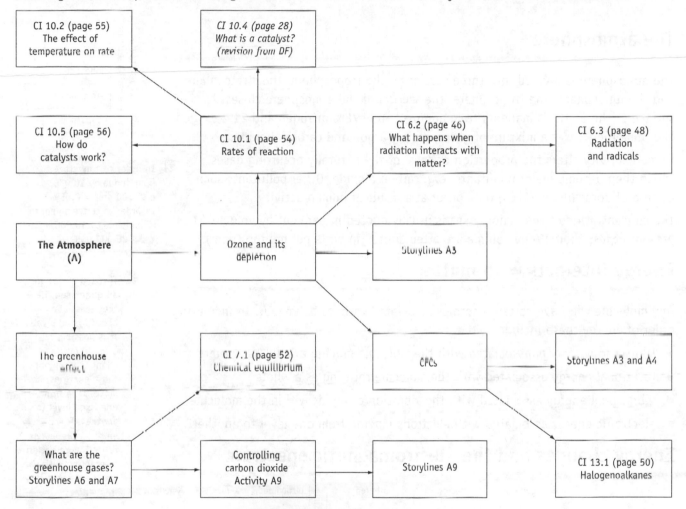

Some sections of the work you need to know do not fit easily into 'Chemical Ideas' and so are added here.

The two most significant 'greenhouse gases' are carbon dioxide and water. High-energy radiation reaches the Earth's surface from the Sun and some frequencies are absorbed. The Earth re-emits lower energy infrared radiation. Greenhouse gases absorb this radiation. This increases the vibrational energy of the molecules of the greenhouse gases and so leads to increased temperatures. We call this **global warming**. Obviously *some* global warming is needed or the Earth would be too cold for us to live on.

CO_2 and H_2O absorb in two bands across the Earth's radiation spectrum. Between these two bands is a 'window' where infrared radiation can escape without being absorbed. About 70% escapes through this window.

CFCs are produced by human activity and are only present in small amounts. However, each molecule has a large 'greenhouse factor' and therefore has a greater contribution to global warming.

In order to reduce global warming the amount of CO_2 put into the atmosphere needs to be reduced. This could be done by reducing our consumption of fossil fuels. Political protocols try to limit the amount of CFCs.

What happens when radiation interacts with matter?

Chemical Ideas 6.2

The atmosphere

The atmosphere is divided into three sections – the troposphere, the stratosphere and the ionosphere. The troposphere, the section of the atmosphere closest to the surface of the Earth, is made up from approximately 78% nitrogen and 21% oxygen. The remaining 1% is a mixture of gases, mainly argon and carbon dioxide.

Human activity alters the proportion of some of the naturally occurring gases, which then become major pollutants, e.g. carbon dioxide. Other pollutants such as chlorofluorocarbons (CFCs) only occur at a result of human activity.

The concentration of the major gases is usually quoted in per cent by volume. Those present at less than 1% by volume are often quoted in parts per million (ppm).

Increasing distance from the Earth's surface
ionosphere
stratosphere
troposphere

> In order to convert % to ppm divide by 10 000, e.g. 500 ppm = 0.05%
> In order to convert ppm to % multiply by 10 000, e.g. 2% = 20 000 ppm

Energy interacts with matter

Any molecule will have certain energies associated with its behaviour. In increasing order of energy these include

- translational energy associated with the molecule moving around as a whole
- rotational energy associated with the molecule rotating as a whole
- vibrational energy associated with the vibration of bonds within the molecule
- electronic energy associated with electrons moving from one level to another.

> These energies are all **quantised**, i.e. they exist at certain fixed levels. It is possible to bring about changes in any of these energies, i.e. move them from one fixed level to another. These energy changes are caused by radiation interacting with the matter at specific frequencies.

Energy changes and the electromagnetic spectrum

	radiofrequency	microwave	infrared	visible	ultraviolet	X-rays	γ-rays

| Frequency/ Hz | 10^5 | 10^6 | 10^7 | 10^8 | 10^9 | 10^{10} | 10^{11} | 10^{12} | 10^{13} | 10^{14} | 10^{15} | 10^{16} | 10^{17} | 10^{18} | 10^{19} | 10^{20} |

| Wavelength/ m | 10^3 | 1 | 10^{-3} | 10^{-6} | 10^{-9} |

Different types of electromagnetic radiation have photons of differing energy associated with them. When the electromagnetic radiation interacts with matter it will bring about changes in the quanta.

increasing energy ↑	Type of change in energy	Type of radiation absorbed
	electronic	ultraviolet or visible
	vibrational	infrared
	rotational	microwave

The energy of photons absorbed to bring about these changes is related to the frequency of the radiation by the equation:

$$E = h\nu$$

> E is the energy associated with one photon (in joules)
> ν is the frequency of the radiation (in Hz)
> h is the Planck constant, 6.63×10^{-34} J Hz^{-1}

Worked example

What is the energy of a photon of blue light with a frequency of 7×10^{14} Hz?

$$E = h\nu = 6.63 \times 10^{-34} \times 7 \times 10^{14} = 4.6 \times 10^{-19} \text{ J}$$

> ◗ Remember the units; this example gives an answer in joules.

Ozone

In the stratosphere dioxygen molecules (O_2) can absorb ultraviolet radiation of the right frequency to split the molecule apart. This is known as **photodissociation**. Oxygen atoms are formed, which are examples of radicals.

> ◗ A radical is a species with one or more unpaired electrons.

$$O_2 \xrightarrow{h\nu} 2O$$

Ozone (O_3) is formed when an oxygen atom (a radical) reacts rapidly with a dioxygen molecule.

$$O + O_2 \rightarrow O_3$$

Ozone is highly reactive. It is destroyed by reacting with radicals present in the stratosphere. If X is a radical the two reactions involved can be written as:

$$X + O_3 \rightarrow XO + O_2$$
$$XO + O \rightarrow X + O_2$$

The 'X' produced in the second equation can then continue the reactions by becoming a reactant in a repeat of the first equation. When the two equations are added together it gives an overall equation

$$O + O_3 \rightarrow O_2 + O_2$$

This equation shows that ozone is being lost. The radical X is involved in the reaction but is not used up and so is acting as a catalyst. We therefore say this is an example of **catalytic recycling**.

> ◗ The radical X could be:
> • HO (the hydroxyl radical), formed from water
> • NO (nitrogen monoxide), produced in internal combustion engines
> • Cl (chlorine radical) produced from the breakdown of CFCs which were used as cleaning solvents, refrigerants or aerosol propellants.

Why is the destruction of ozone a problem?

Ozone absorbs radiation in the region 10.1×10^{14} Hz to 14.0×10^{14} Hz. This is the ultraviolet region of the electromagnetic spectrum and is most damaging to the skin. Because much of this UV radiation is absorbed by ozone in the stratosphere damage to the skin, such as skin cancer, is reduced.

In areas of the stratosphere where ozone has been depleted UV radiation can pass through to the troposphere and the risk of skin damage can increase.

> **?** *Quick check questions*
>
> 1 (a) What is the energy of a photon of frequency 3.8×10^{13} Hz?
>   (b) To which type of radiation does this correspond?
>   (c) The change between which type of energy levels is brought about by this radiation?
>
> 2 (a) Write out balanced equations to show how a chlorine radical could be involved in the removal of ozone from the atmosphere.
>   (b) Why is this an example of catalytic recycling?

Radiation and radicals
Chemical Ideas 6.3 and Chemical Storylines A4

In a covalent bond a pair of electrons is shared between two atoms. When the bond breaks the electron pair is redistributed. This can occur in one of two ways: heterolytic fission or homolytic fission.

- In heterolytic fission both the electrons of the shared pair go to just one of the atoms when the bond breaks. This forms ions.

$$H : Cl \longrightarrow H^+ + Cl^-$$

- In homolytic fission one of the two electrons in the shared pair goes to each of the atoms. Both atoms now have one unpaired electron.

$$Br : Br \longrightarrow 2Br^{\bullet}$$

If the radical formed has two unpaired electrons it is called a biradical. An example is O_2.

$$\bullet O{-}O \bullet$$

As a result of the unpaired electrons radicals are very reactive.

> A radical is a species with one or more unpaired electrons.

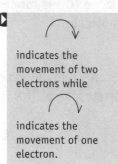

> indicates the movement of two electrons while
>
> indicates the movement of one electron.

Mechanism of radical chain reaction

Radicals are highly reactive and undergo chain reactions. Chain reactions can be divided into three stages – initiation, propagation and termination. Be sure you can identify which stage an equation represents.

Initiation   There are no radicals at the beginning of this stage but radicals are formed by the end of the stage:

$$Cl \overset{\times}{\underset{}{Cl}} \xrightarrow{h\upsilon} 2Cl^{\bullet}$$

Propagation   There are radicals at the start of this stage of the reaction and new radicals are formed by the end of the stage.

$$Cl^{\bullet} + H \overset{\times}{\underset{}{H}} \longrightarrow Cl{-}H + H^{\bullet}$$

$$H^{\bullet} + Cl \overset{\times}{\underset{}{Cl}} \longrightarrow H{-}Cl + Cl^{\bullet}$$

Termination   The reaction is terminated when two radicals collide:

$$\text{e.g. } H^{\bullet} + H^{\bullet} \longrightarrow H{-}H$$

> In this equation $h\upsilon$ represents energy.

> Radical reactions are fast, are often initiated by heat or light and normally occur in the gas phase.

The reactions of alkanes with halogens

A halogen can substitute a hydrogen on an alkane chain, via a radical substitution mechanism, to produce a halogenoalkane.

Initiation   Homolytic fission occurs in the presence of UV light: $Cl_2 \xrightarrow{h\upsilon} 2Cl^{\bullet}$

Propagation

$$H{-}\overset{\overset{\displaystyle H}{|}}{\underset{\underset{\displaystyle H}{|}}{C}}{-}H + Cl^{\bullet} \longrightarrow CH_3^{\bullet} + HCl \qquad CH_3^{\bullet} + Cl_2 \longrightarrow CH_3Cl + Cl^{\bullet}$$

Termination

$$CH_3^{\bullet} + CH_3^{\bullet} \longrightarrow C_2H_6 \qquad CH_3^{\bullet} + Cl^{\bullet} \longrightarrow CH_3Cl$$

> When radicals form due to the presence of light, this is known as **photodissociation**.

Depletion of ozone (Chemical Storylines A4)

Ozone is removed in a catalytic cycle, as discussed in Chemical Ideas 6.2 (see page 47). One single chlorine atom can remove about one million ozone molecules. These chlorine atoms are produced by the breakdown of chlorofluorocarbons (CFCs).

CCl_2F_2 was the CFC which Midgley first developed to replace ammonia as a refrigerant. CFCs are also used as propellants in aerosols, blowing agents for making expanded plastics and as cleaning solvents. CFCs are good at these jobs because they are unreactive, have a range of conveniently low boiling points, have low toxicity and are very stable.

It is this last property that causes the problem. CFCs are estimated to have a lifetime in the troposphere of approximately 100 years. When they reach the stratosphere they undergo photodissociation to produce chlorine radicals, which then go on to remove ozone.

Alternatives to CFCs

Replacements have been used for CFCs, although they in turn have their own problems.

Replacement	Advantages	Disadvantages
Hydrochlorofluorocarbons (HCFCs) and hydrofluorocarbons (HFCs)	H–C bonds are broken down in the troposphere before the compounds have chance to reach the stratosphere	Are greenhouse gases that contribute to global warming
Alkanes	Alkanes don't contain chlorine	Flammable, are greenhouse gases

Why is it important to prevent ozone depletion?

You need to note that at tropospheric levels (i.e. near the Earth's surface) ozone is a troublesome gas that is harmful to human health, weakening the immune system.

Ozone is formed in the stratosphere by the following reactions:

$$O_2 \xrightarrow{h\upsilon} 2O^\bullet \quad \text{then} \quad O^\bullet + O_2 \rightarrow O_3$$

In the stratosphere ozone absorbs UV radiation. This prevents these harmful frequencies reaching the Earth's surface. For humans this is good. UV radiation is linked to skin damage and, in extreme cases, skin cancer. If ozone levels are reduced this could lead to increased numbers of cases of skin cancer.

? Quick check questions

1 What is the difference in the first products of heterolytic fission and homolytic fission?
2 What are the three stages of a radical chain reaction?
3 What is the difference between a radical and a nucleophile? (see page 51)

Halogenoalkanes

Chemical Ideas 13.1

The homologous series of the halogenoalkanes is an alkane series with hydrogen atoms substituted by one or more halogen atoms. They are often shown as R–Hal, where Hal could be F, Cl, Br or I.

> Halogenoalkanes are sometimes called haloalkanes.

Naming halogenoalkanes

The alkane chain name is *prefixed* with the name of the halogen. The halogens are listed in alphabetical order, with a number indicating the position of each.

This is
3-bromo-1-chlorobutane.

> Each halogen atom is prefixed with a number, e.g. 1,1-dichloro-2-iodopropane.
>
> There are two chlorine atoms attached to carbon 1 so there are two numbers here.

Physical properties of halogenoalkanes

The boiling points increase with a heavier halogen atom (R–I > R–F) or with increasing numbers of halogen atoms (CCl_4 > CH_2Cl_2). As the halogen introduced is larger or the number of halogen atoms increases, the overall number of electrons increases. This strengthens the instantaneous dipole–induced dipole interactions (see page 60). With greater intermolecular forces more energy is needed to pull the molecules apart (as in boiling) so the boiling point is higher.

Bond enthalpies and reactivity of halogenoalkanes

Bond within the molecule	Bond strength	Reactivity
C–F C–Cl C–Br C–I	decreasing strength ↓	increasing reactivity ↓

The C–Hal bond becomes weaker as the size of the halogen increases. This makes the bond easier to break and the compounds become more reactive.

- Fluoro- compounds are very unreactive.
- Chloro- compounds are reasonably stable in the stratosphere and can react to produce chlorine radicals that deplete ozone.
- Bromo- and iodo- compounds are reactive so are useful as intermediates in chemical synthesis.

Reactions of halogenoalkanes

1. Homolytic fission

$$R\!-\!Cl \xrightarrow{h\upsilon} R^{\bullet} + Cl^{\bullet}$$

> Conditions: gas phase with high temperatures or the presence of UV radiation (e.g. in the stratosphere).

2. Heterolytic fission $R–Hal \rightarrow R^+ + Hal^-$

The carbon–halogen bond breaks to give ions. If the polar C–Hal bond is broken completely a negative halide ion moves away, leaving the C group behind positively charged. This is now a **carbocation**.

$$CH_3-\underset{\underset{CH_3}{|}}{\overset{\overset{CH_3}{|}}{C}}\text{—}Cl \longrightarrow CH_3-\underset{\underset{CH_3}{|}}{\overset{\overset{CH_3}{|}}{C^+}} + Cl^-$$

> ▶ Conditions: dissolved in a polar solvent such as an ethanol/water mix.

3. Substitution reactions $R–Hal + X^- \rightarrow R–X + Hal^-$

In this case the C–Hal bond breaks and the halogen atom is replaced by another group. Since the halogen is replaced by a nucleophile these reactions are called **nucleophilic substitution** reactions.

> ▶ X^- is an example of a nucleophile.

Worked example

Mechanism for the nucleophilic substitution reactions of halogenoalkanes.

$$\underset{\text{nucleophile}}{\overset{..}{X}^-} \curvearrowright \overset{\delta+}{\underset{}{\diagdown}}C\overset{\delta-}{\text{—}}Hal \longrightarrow \diagdown C\text{—}X + Hal^-$$

> ▶ Nucleophiles have one or more lone pairs of electrons that they can donate to form new bonds. Examples include:
>
> $H-\overset{..}{\underset{..}{O}}:^-$ hydroxide ion
>
> $H\overset{\overset{..}{O}..}{\diagup \diagdown}H$ water
>
> $\underset{H \; H \; H}{\overset{N}{|}}$ ammonia

Step 1 The δ+ carbon from the carbon–halogen bond is attacked by the nucleophile.

Step 2 The lone pair of electrons on the nucleophile form a new bond with the carbon from the carbon–halogen bond.

Step 3 At the same time the carbon–halogen bond breaks, giving a halide ion.

> ▶ A full headed arrow indicates the movement of a pair of electrons.

Nucleophile	Equation	Product	Reaction conditions
HOH	$R–Hal + H_2O \rightarrow R–OH + H^+ + Hal^-$	alcohol	heat under reflux: this is sometimes called hydrolysis
OH^-	$R–Hal + OH^- \rightarrow R–OH + Hal^-$	alcohol	heated under reflux with NaOH(aq) with ethanol as solvent
NH_3	$R–Hal + NH_3 \rightarrow R–NH_2 + Hal^- + H^+$	amine	the halogenoalkane is heated with concentrated ammonia solution in a sealed tube

> ▶ You will need to learn the reaction conditions for these reactions.

Preparation of halogenoalkanes

You need to know the practical details for the preparation of a halogenoalkane (see Activity A4.2 for details). This preparation involves a nucleophilic substitution reaction of a protonated alcohol. The reaction is carried out under reflux and the liquid mixture is purified using distillation.

The mechanism is as follows:

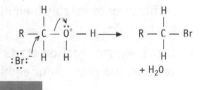

? **Quick check questions**

1 Draw the skeletal formula for the following compound and name it.

$$CH_3CH_2CHBrCCl_2CHICH_3$$

2 What are the conditions needed for ammonia to react with a halogenoalkane?

3 Why is iodoethane more reactive than fluoroethane?

Chemical equilibrium
Chemical Ideas 7.1

Dynamic equilibrium

A chemical reaction has a forward reaction, but it also has a backward reaction. If the backward reaction is significant then the reaction is reversible. When the rate of the forward reaction is the same as the rate of the backward reaction a system is said to be in **dynamic equilibrium**. This is represented by the symbol $\rightleftharpoons$

For example, when carbon dioxide dissolves in water the following reversible changes occur: $CO_2(g) \rightleftharpoons CO_2(aq)$ and $H_2O(l) \rightleftharpoons H_2O(g)$

These are physical changes. The following reversible chemical change also occurs:

$$CO_2(aq) + H_2O(l) \rightleftharpoons HCO_3^-(aq) + H^+(aq)$$

Hydrogencarbonate ions (HCO_3^-) and hydrogen ions (H^+) form.

Steady state

Strictly speaking a chemical equilibrium can only be established in a **closed system**. In an open system a series of reactions can only come to a steady state. An example of a steady state is the production and destruction of ozone in the stratosphere.

$$O + O_2 \longrightarrow O_3 \qquad \text{ozone production}$$
$$O_3 \xrightarrow{hv} O_2 + O \qquad \text{ozone destruction}$$
$$O + O_3 \longrightarrow O_2 + O_2 \quad \text{overall}$$

None of these reactions comes to equilibrium, but left to themselves they will reach a point when ozone is being produced as fast as it is being used up. The series has reached a **steady state**.

Position of equilibrium

For any given reversible reaction there are many combinations of equilibrium mixtures possible. These combinations depend on the original concentrations of the substances and the conditions. We use the term **position of equilibrium** to describe one set of equilibrium concentrations for a reaction.

- If most of the reactants become products before the reverse reaction increases sufficiently to establish equilibrium we say that the position of equilibrium lies to the right.

- If the reverse reaction starts to increase as soon as some products are formed we say that the position of equilibrium lies to the left.

Le Chatelier's principle

The position of the equilibrium can be altered by changing the concentration of solutions, the pressure of gases or the temperature.

> Le Chatelier's principle states that if a system is at equilibrium, and a change is made in any of the conditions, then the system responds to counteract the change as much as possible.

A catalyst does not change the position of equilibrium, just the rate at which the equilibrium is established.

Concentration

Concentration change	Equilibrium shift
increasing reactants	to the right (decreases reactants)
increasing products	to the left (decreases products)
decreasing reactants	to the left (increases reactants)
decreasing products	to the right (increases products)

$$CaCO_3(s) \rightleftharpoons CaO(s) + CO_2(g)$$

In the production of calcium oxide from calcium carbonate, the carbon dioxide is removed from the kiln in order to encourage the position of equilibrium to move to the right and increase the yield of calcium oxide.

Pressure

Pressure change	Equilibrium shift	Example
increasing	to the side with fewer molecules (in this case, to the right)	$CO(g) + 2H_2(g) \rightleftharpoons CH_3OH(g)$ 3 molecules 1 molecule
decreasing	to the side with more gas molecules (in this case, to the right)	$CH_4(g) + H_2O(g) \rightleftharpoons CO(g) + 3H_2(g)$ 2 molecules 4 molecules

Temperature

Temperature change	Equilibrium shift	Example
increase	position of equilibrium shifts in the direction of the endothermic reaction (in this example the equilibrium mixture becomes darker brown)	$$2NO_2(g) \underset{endothermic}{\overset{exothermic}{\rightleftharpoons}} N_2O_4(g)$$ brown gas colourless gas
decrease	position of equilibrium shifts in the direction of the exothermic reaction (in this example the equilibrium mixture becomes lighter brown)	

> ## ? Quick check questions
>
> 1 $C_2H_4(g) + H_2O(g) \rightleftharpoons C_2H_5OH(g)$; $\Delta H = -46$ kJ mol^{-1}, $P = 70$ atm, $T = 300°C$
>
> Predict, using Le Chatelier's principle, the changes in (a) concentration, (b) temperature and (c) pressure that would move the position of equilibrium to the right?
>
> 2 $Fe^{3+}(aq)$ $+$ $SCN^-(aq)$ $\rightleftharpoons$ $[FeSCN]^{2+}(aq)$
>   pale yellow   colourless   blood red
>
> For this reaction, what would you see if you added (a) more $Fe^{3+}(aq)$ and (b) more $[FeSCN]^{2+}(aq)$ to the reaction mixture?

Rates of reaction

Chemical Ideas 10.1

Rates of reaction can be affected by a number of factors:

- concentration
- intensity of radiation
- surface area
- pressure
- particle size
- a catalyst
- temperature

> ◖ Reaction kinetics is the study of the rate of a reaction.

Collision theory

Reactions occur when particles of reactants collide with a certain minimum kinetic energy:

- at higher concentrations and higher pressures the particles are in closer proximity to each other, encouraging more frequent collisions
- at higher temperatures a much higher proportion of colliding particles have sufficient energy to react and more particles are able to overcome the activation enthalpy barrier
- with smaller pieces of reactant there is a larger surface area on which the reactions can take place, so the greater the chance of a successful collision
- heterogeneous catalysts provide a surface where reacting particles may break and make bonds.

Activation enthalpy

This is the minimum kinetic energy required by a pair of colliding particles before a reaction will occur.

An enthalpy profile shows how the enthalpy changes as a reaction proceeds.

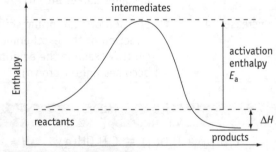

? ## Quick check questions

1 Using collision theory, explain why the rate of ozone depletion has increased as more CFCs have been released into the atmosphere.

2 (a) Which enthalpy profile has the highest activation enthalpy **I** or **II**?

   (b) Which reaction is endothermic?

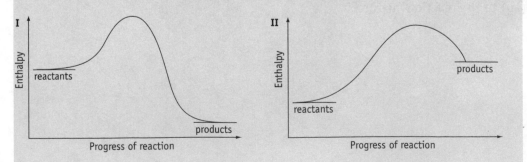

The effect of temperature on rate

Chemical Ideas 10.2

Rates of reaction do not just depend on how frequently particles collide, but on how much energy they have when a collision takes place.

Collision theory states that reactions occur when molecules collide with a certain *minimum* kinetic energy. This minimum kinetic energy is called the **activation enthalpy**. The energy needed to overcome the energy barrier is called the activation energy barrier.

As the temperature increases, the rate of a chemical reaction also increases. The reason for this is because of the distribution of energies amongst the reacting particles. This distribution is called the **Maxwell–Boltzmann distribution**.

There needs to be enough molecules with sufficient energy for a reaction to take place. The molecules need to have kinetic energy higher than the enthalpy of activation.

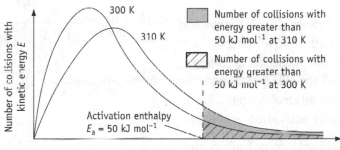

Reactions go faster at higher temperatures because a larger proportion of the colliding particles have the minimum activation enthalpy needed to react.

Look back at the diagram above:

- the peak of the number of collisions at 300 K is at a lower kinetic energy value than the peak at 310 K
- at the kinetic energy value of 50 kJ mol^{-1} the number of collisions at 310 K is almost twice as many as the number of collisions at 300 K
- in this case when the temperature rises by 10°C (10 K) the rate of reaction approximately doubles.

? Quick check questions

1 What does E_a represent?

2 Shade the area of the graph representing the number of collisions with sufficient energy to lead to a reaction.

3 On the graph sketch a line showing the energy distribution for collisions occurring at $(T_1 + 10)$ K. Label this curve T_2.

4 Shade the area (differently) representing the number of collisions with sufficient energy to lead to a reaction at T_2 K.

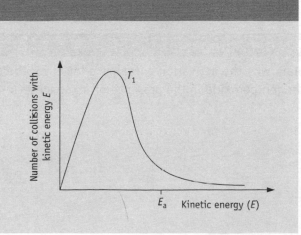

How do catalysts work?
Chemical Ideas 10.5

In order for any chemical reaction to proceed, bonds need first to be broken before the new bonds can be made. Bond breaking is an endothermic process and energy is taken in to break the bonds.

A pair of reacting molecules must have more energy than the activation energy value in order to make a successful collision. Catalysts are used in order to overcome the activation energy barrier more easily. With a catalyst, successful collisions can take place at a lower energy. This is called lowering the activation energy barrier.

The reaction profiles for a catalysed reaction and an uncatalysed reaction are shown here.

Catalysts work by providing an alternative reaction pathway for the breaking and remaking of bonds. This alternative path has a lower activation enthalpy.

Catalysts do not affect the position of equilibrium in a reversible reaction.

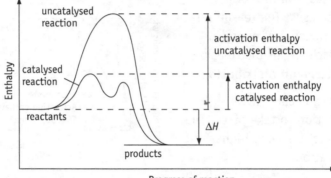

Heterogeneous catalysts provide a surface on which a reaction may take place, thus lowering the energy needed for a successful collision. This thus lowers the activation energy barrier.

Homogenous catalysts work by forming an intermediate compound with the reactants. The diagram above has two humps on the lower (catalysed) pathway, one for each step.

- In the first step the activation energy barrier is overcome and an intermediate is formed.
- In the second step this intermediate breaks down to give a product and re-form the catalyst.

? Quick check question

1 Explain why the activation enthalpy is lower for the decomposition of hydrogen peroxide in the presence of manganese(IV) oxide.

The Polymer Revolution (PR)

Many chance discoveries led to the development of the polymers we take for granted in our everyday lives. This unit looks at addition polymerisation. You will cover the following themes. CI refers to sections in your Chemical Ideas textbook.

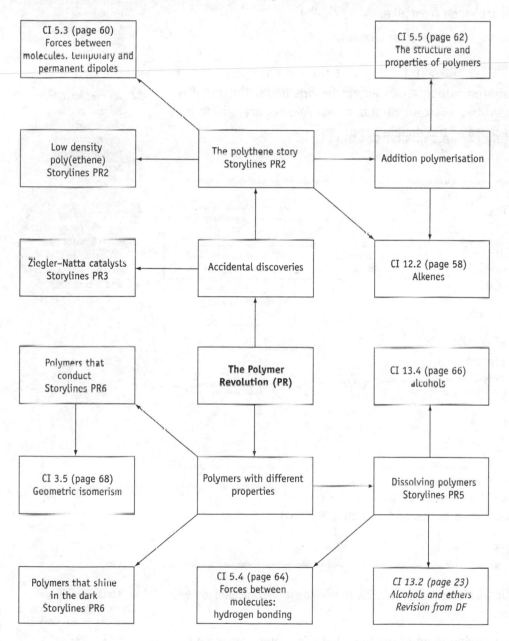

CI 5.3 (page 60)
Forces between molecules. Temporary and permanent dipoles

CI 5.5 (page 62)
The structure and properties of polymers

Low density poly(ethene)
Storylines PR2

The polythene story
Storylines PR2

Addition polymerisation

Ziegler–Natta catalysts
Storylines PR3

Accidental discoveries

CI 12.2 (page 58)
Alkenes

Polymers that conduct
Storylines PR6

The Polymer Revolution (PR)

CI 13.4 (page 66)
alcohols

CI 3.5 (page 68)
Geometric isomerism

Polymers with different properties

Dissolving polymers
Storylines PR5

Polymers that shine in the dark
Storylines PR6

CI 5.4 (page 64)
Forces between molecules:
hydrogen bonding

CI 13.2 (page 23)
Alcohols and ethers
Revision from DF

Alkenes

Chemical Ideas 12.2

Alkenes are the basic hydrocarbon units of many polymers:

- –A–A–A–A–A–A– polymers are made from alkenes
- –A–B–A–B–A–B– polymers *are not* made from alkenes.

The general formula of alkenes is C_nH_{2n}.

Alkenes have single bonds between carbon atoms except for one bond. This bond is a double bond and it is between two adjacent carbon atoms. Alkenes are said to be **unsaturated** hydrocarbons because of the double bond.

> Hydrocarbons are molecules made of carbon and hydrogen <u>only</u>.

> Alk**anes** have just carbon–carbon single bonds; alk**enes** have a carbon–carbon double bond.

Naming alkenes

Alkene names end in '–ene'. The number preceding the '–ene' indicates the position of the double bond. But-1-ene is $CH_3-CH_2-CH=CH_2$ and but-2-ene is $CH_3-CH=CH-CH_3$.

Alkenes can form cyclic compounds.

An alkene with two double C=C bonds is called a diene. Note the addition of the letter 'a' after 'hex' in this example: hexa-1,3-diene is $CH_2=CH-CH=CH-CH_2-CH_3$

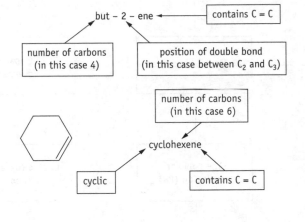

> When naming organic compounds always put a dash (-) between a number and a letter and a comma between numbers.

Shape of alkenes

All bond angles are 120° since there are three groups of electrons around each carbon atom (two single bonds and one double bond).

Reactions of alkenes

Alkenes undergo **electrophilic addition** reactions. In the following examples we will use ethene as a typical alkene.

There are four electrons in the double bond of ethene. These four electrons give the region between the two carbon atoms a high density of negative charge.

Electrophiles are attracted to this negatively charged region in an alkene and accept a pair of electrons from the double bond in a reaction.

> **Electrophiles** are either positive ions or molecules with a partial positive charge on one of the atoms.

A general scheme for the reactions is:

You need to know the mechanism for the addition of bromine to an alkene
(i.e. where X–Y is Br–Br).

- When a bromine molecule approaches an alkene it becomes **polarised**.
- The electrons in the bromine molecule are repelled back along the molecule.
- The electron density is **unequally distributed**.
- The bromine atom nearest to the alkene becomes slightly positively charged.
- It now acts as an **electrophile**. A pair of electrons moves towards the slightly charged bromine atom. A C–Br bond is formed.
- The carbon species is now positively charged. It is a **carbocation**.
- The other bromine, now negatively charged, moves in rapidly to make another bond.

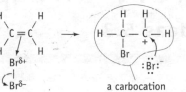

the bromine molecule becomes polarised

a carbocation

> The process is addition by an electrophile across a double bond. It is **electrophilic addition**.

Alkenes can react with a number of electrophiles. Make sure you know the reaction conditions for the following:

Electrophile	Product	Conditions
Br_2	CH_2BrCH_2Br 1,2-dibromoethane	room temperature, no catalyst
$Br_2(aq)$	CH_2BrCH_2OH 2-bromoethan-1-ol	room temperature
HBr	CH_3CH_2Br bromoethane	aqueous solution, room temperature
H_2O (H–OH)	CH_3CH_2OH ethanol	phosphoric acid/silica, 300°C/60 atm or 1 atm with conc. H_2SO_4
H_2	CH_3CH_3 ethane	Pt catalyst, room temperature and pressure or Ni catalyst, 150°C/5 atm

> The products are different for Br_2 as a liquid and $Br_2(aq)$ since in water the water can attack the intermediate carbocation.

Addition polymerisation

Under the right conditions, alkenes can undergo addition polymerisation. The small unsaturated starting molecules are called **monomers** and they join together to form a long chain saturated **polymer**. The mechanism is **free radical**.

$CH_2 = CH_2 + CH_2 = CH_2 + CH_2 = CH_2 \longrightarrow -CH_2 - CH_2 - CH_2 - CH_2 - CH_2 - CH_2 -$

ethene (monomer)

poly(ethene) (polymer)

This may be written as:

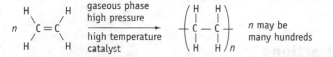

> The polymer is named by putting the name of the monomer in brackets and prefixing with poly; e.g. choroethene monomer gives poly(chloroethene) as the polymer. Note, however, that the polymer is <u>not</u> an alkene.

❓ *Quick check questions*

1. What is the structural formula of pent-1-ene?
2. Draw a skeletal formula of the product when $H_2(g)$ reacts with pent-2-ene.
3. Draw out the structure of the polymer formed when the monomer CHCl=CHCl undergoes addition polymerisation.

Forces between molecules: temporary and permanent dipoles

Chemical Ideas 5.3

In any liquid or solid there are forces between molecules. These are called **intermolecular forces**.

Polar molecules

A **dipole** is a molecule (or part of a molecule) with a positive end and a negative end.

When a molecule has a dipole we say it is **polarised**. Molecules with a **permanent** dipole are **polar** molecules.

▶ For example:
$$\delta+ \quad \delta-$$
$$H - Br$$

Temporary, or instantaneous, dipoles

If a molecule does not have a permanent dipole the electron density in the molecule may be unevenly distributed at any one time. It has an **instantaneous** dipole. The swirling electron density changes distribution and so the polarity will change.

Electron cloud evenly distributed; no dipole.

$$\delta+ \qquad \delta-$$

At some instant, more of the electron cloud happens to be at one end of the molecule than the other; molecule has an instantaneous dipole.

If other molecules are nearby they may cause an effect and produce an **induced** dipole.

An unpolarised Cl_2 molecule finds itself next to an HCl molecule with a permanent dipole.

Electrons get attracted to the positive end of the HCl dipole, inducing a dipole in the Cl_2 molecule.

Intermolecular forces

Molecular substances which contain dipoles attract each other. Two of these kinds of attraction are instantaneous dipole–induced dipole attractions and permanent dipole–permanent dipole attractions.

1. Instantaneous dipole–induced dipole attractions

These are the weakest type of intermolecular attraction. They can happen in *all types of molecule*, even those with a permanent dipole already.

Consider krypton atoms. Their electrons are continually moving, creating instantaneous dipoles. As other krypton atoms approach an atom with an instantaneous dipole they will produce an induced dipole. This is an **instantaneous dipole–induced dipole** attraction. Obviously because the electron cloud in the krypton atoms is continually

▶ Remember, in order for a substance to melt or boil intermolecular forces need to be broken. The greater these intermolecular forces the more energy is needed to break them so the higher the melting or boiling point.

moving these instantaneous dipole–induced dipole interactions are continually forming and breaking. The more electrons an atom or molecule has the greater these attractions and the higher the boiling point of the substance.

The only intermolecular forces between chains of poly(ethene) are instantaneous dipole–induced dipoles yet poly(ethene) is solid at room temperature. This is because the chains are long and can pack closely together. This means although the intermolecular forces are very weak there are a lot of them.

2. Permanent dipole–permanent dipole attractions

Molecules with permanent dipoles have atoms with different **electronegativity** values. The slightly positively charged end of a molecule attracts the slightly negatively charged end of the next molecule and an intermolecular attraction occurs. This is stronger than the attraction noted between noble gas atoms. Since the dipoles in both molecules are permanent this type of intermolecular attraction is called a **permanent dipole–permanent dipole** attraction.

$$\overset{\delta+}{} \overset{\delta-}{} \quad \overset{\delta+}{} \overset{\delta-}{} \quad \overset{\delta+}{} \overset{\delta-}{}$$
$$--- H — Br--- H — Br--- H — Br$$

Permanent dipole–permanent dipole attractions are stronger than instantaneous dipole–induced dipole attractions but are weaker than hydrogen bonds (see page 64).

Permanent dipole–permanent dipole attractions are responsible for holding polyester molecules together.

Polyester molecules

Electronegativity

The degree to which an atom of an element attracts electrons is called its electronegativity. The more electronegative an element is the greater its attraction for electrons.

$$F > O > Cl > Br > I > S > C > H$$

Note that the difference between the electronegativities of C and H is so small that bonds between them can be considered to be non-polar.

Bond polarity and polar molecules

A polar molecule is one that has a permanent dipole, for example ethanoic acid:

If the differences in electronegativities of the elements in a molecule are very small the dipole is negligible, for example in CH_4. Sometimes even if the bonds are polar a molecule might not have a dipole. This is due to the arrangement of the atoms in the molecule. For example, in tetrachloromethane each C–Cl bond is polar, but the symmetrical arrangement of the chlorine atoms means there is no overall dipole.

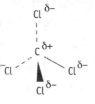

> You may be asked to identify the strongest intermolecular force in a given molecule and label it on a diagram. If so, always label the bond polarities and identify the attraction between the polarised bonds, as in the diagram above.

Quick check questions

1 Would you expect xenon or krypton to have the higher boiling point? Explain your answer.

2 Why do branched chain hydrocarbons have lower boiling points than straight chain hydrocarbons of the same relative formula mass?

Structure and properties of polymers

Chemical Ideas 5.5 (revisited in the A2 course)

A **polymer** is a long molecule made up from lots of small molecules. The small molecules that add together to form a polymer are called **monomers**.

Addition polymerisation

If the monomers contain a double bond they can add together to make a polymer:

$$A + A + A + A \rightarrow -A-A-A-A-$$

This is called **addition polymerisation**.

Co-polymerisation

A co-polymer occurs when two different monomers become incorporated into the final polymer, in this example propene and ethene have co-polymerised.

$$-CH_2-CH-CH_2-CH-CH_2-CH_2-CH_2-CH-CH_2-CH- \atop \qquad | \qquad\quad | \qquad\qquad\qquad\qquad | \qquad\quad |$$
$$\qquad CH_3 \quad\; CH_3 \qquad\qquad\qquad CH_3 \;\; CH_3$$

Properties of polymers

The properties of a polymer are dependent on six factors. These are discussed below.

Chain length   The longer the chains the stronger the polymer. Tensile strength is a measure of how much force needs to be applied before a polymer snaps. Tensile strength increases with increasing chain length because:

- longer chains become more entangled
- longer chains have greater intermolecular forces between them and so are more difficult to pull apart.

Side groups on the polymer chain   The more polar the side groups (such as Br or Cl) the stronger the attraction between polymer chains, therefore the stronger the polymer.

Branching   The more unbranched the chain the closer the packing, so the stronger the attraction between the chains. This makes the polymer stronger.

Chain flexibility   The more rigid the chain the stronger the polymer.

Cross-linking   The more extensive the cross-linking the harder it is to melt the polymer.

Stereoregularity   The more regular the orientation of the side groups, the closer the packing, the stronger the polymer.

Thermoplastics

These are polymers without cross-links between the chains. The intermolecular forces between the chains are much weaker than the covalent cross-links in a thermoset (see below).

The attractive forces in thermoplastics can be broken by warming. The chains can move over each other and the polymer can be deformed, i.e. change shape. On cooling the weak forces between the polymer chains re-form and the

Thermoplastic: no cross–linking

HEAT

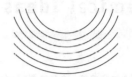

Weak forces between polymer chains easily broken by heating; polymer can be moulded into new shape.

thermoplastic holds its new shape. Examples include poly(ethene) and nylon.

Thermosets

These polymers have extensive cross-linking between the different polymer chains. The forces between the chains are much stronger than in thermoplastics.

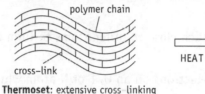

polymer chain

cross–link

Thermoset: extensive cross-linking

HEAT

Strong covalent bonds between polymer chains cannot be easily broken; polymer keeps shape on heating.

The attractive forces cannot be broken by warming. The chains cannot move relative to each other and the polymer cannot change shape. If heating continues the polymer just chars and burns. An example of this type of polymer is Bakelite.

Crystallinity

The areas where the chains are closely packed in a regular way are called crystalline. The higher the order the more crystalline. The higher the crystallinity, the stronger and less flexible the polymer. This is because in areas of high crystallinity the polymer chains pack closer together so the intermolecular forces are greater. This makes it more difficult for the chains to slide over each other. Areas of a polymer where crystallinity is low are called amorphous regions.

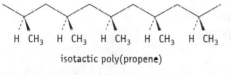

isotactic poly(propene)

Regular structure: crystalline
Strong and rigid – like hdpe

Used in sheet and film form for packaging and containers; used to make fibres for carpets

atactic poly(propene)

Irregular structure: amorphous
Soft and flexible

Used to make roofing materials, sealants and other weatherproof coatings

Quick check questions

1 What is the difference between a thermoset and a thermoplastic polymer?

2 Give three ways in which a polymer may be made stronger and less flexible.

3 Explain what is meant by addition polymerisation.

Forces between molecules: hydrogen bonding

Chemical Ideas 5.4

Hydrogen bonding

Hydrogen bonds are much stronger than other types of intermolecular forces, e.g. instantaneous dipole–induced dipole forces.

Hydrogen bonds have the following features:

- there is a large dipole between a small hydrogen atom and a highly electronegative atom (such as O)
- the small H atom is able to approach close to other atoms to form the hydrogen bond
- there needs to be a lone pair of electrons on an O, F or N atom which the hydrogen can line up with.

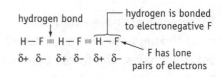

The water molecule

The water molecule consists of two hydrogen atoms covalently bonded to one oxygen atom. Thus the oxygen atom has two bond pairs and two lone pairs of electrons.

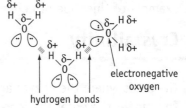

Each water molecules can form twice as many hydrogen bonds as hydrogen fluoride. Water is unique in this respect and hydrogen bonding is the cause of some of water's unusual properties.

Hydrogen bonding in water and ice

Water has a high specific heat capacity. A lot of energy is required to overcome the hydrogen-bonded water molecule clusters. Therefore the energy absorbed is not used to increase the kinetic energy of the molecules and thus raise the temperature. Instead it is stored for longer. This allows the oceans to act as a temperature regulator.

For most substances, as the temperature drops below its freezing point the density increases as a solid forms. This is not the case for water. The density of ice changes at around 4°C. In the diagram (on the next page) there are four groups around each O atom, arranged tetrahedrally in three dimensions by hydrogen bonding. This leads to an 'open' structure with large spaces in it. Therefore the density of ice is less than the density of water at 273 K (0°C).

This means that ice will float on ponds, keeping fish alive in the insulated liquid water underneath, but the expansion in uninsulated water pipes causes cracks and leaks.

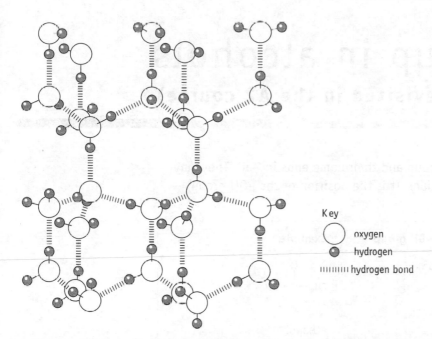

A diagram of the open structure of ice.

Dissolving polymers

The reason why polymers such as poly(ethanol) dissolve depends on their structure. Poly(ethanol) has the ability to form hydrogen bonds.

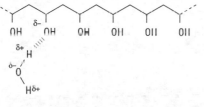

Hydrogen bonding between poly(ethanol) and water.

The –OH groups on the chain can hydrogen bond with water molecules, so the polymer is soluble. The solubility can be changed by altering the proportion of –OH groups in the polymer. The polymer will become less water soluble as the proportion of –OH groups decreases. If the proportion is too large the molecule will undergo a large degree of internal hydrogen bonding and it will take so much energy to pull the molecules apart that the polymer becomes virtually insoluble in water. These properties make poly(ethanol) useful for making dissolving laundry bags for use in hospitals, dissolving seed coatings and dissolving stitches for use in surgery.

? *Quick check questions*

1   Which atoms have high enough electronegativity to form a hydrogen bond with hydrogen atoms?

2   Why is ammonia highly soluble in water?

3   Explain why the density of water at 0°C is less than expected.

The –OH group in alcohols
Chemical Ideas 13.4 (revisited in the A2 course)

Alcohols all contain the –OH functional group and their name ends in '-ol'. They may be classified as primary, secondary or tertiary. It is the position of the –OH group which determines the classification.

Type of alcohol	Position of –OH group	Example
Primary	at end of chain $R - \overset{\displaystyle H}{\underset{\displaystyle H}{\overset{\textstyle \mid}{\underset{\textstyle \mid}{C}}}} - OH$	$CH_3 - \overset{\displaystyle H}{\underset{\displaystyle H}{\overset{\textstyle \mid}{\underset{\textstyle \mid}{C}}}} - OH$ ethanol
Secondary	in middle of chain $R^1 - \overset{\displaystyle H}{\underset{\displaystyle OH}{\overset{\textstyle \mid}{\underset{\textstyle \mid}{C}}}} - R^2$	$CH_3CH_2 - \overset{\displaystyle H}{\underset{\displaystyle OH}{\overset{\textstyle \mid}{\underset{\textstyle \mid}{C}}}} - CH_3$ butan-2-ol
Tertiary	at junction of chains $R^1 - \overset{\displaystyle R^2}{\underset{\displaystyle OH}{\overset{\textstyle \mid}{\underset{\textstyle \mid}{C}}}} - R^3$	$CH_3CH_2 - \overset{\displaystyle CH_3}{\underset{\displaystyle OH}{\overset{\textstyle \mid}{\underset{\textstyle \mid}{C}}}} - CH_3$ 2-methylbutan-2-ol

Oxidation of alcohols

The –OH group can be oxidised by using acidified potassium dichromate(VI) ($K_2Cr_2O_7$). This is due to the oxidation of the –OH group to a carbonyl group.

At the same time the $Cr_2O_7^{2-}$(aq) ion (which is orange) is reduced to Cr^{3+}(aq) (which is green).

The reaction conditions for the oxidation of alcohols are as follows: heating the alcohol under reflux with excess acidified potassium dichromate(VI) solution.

The products of oxidation

The product depends on the type of alcohol used.

With a primary alcohol an aldehyde is produced, which can oxidise further to give a carboxylic acid. The colour of the reaction mixture changes from orange to green.

If the aldehyde is required it can be distilled out of the reaction mixture as it is produced, in order to prevent further oxidation.

With a secondary alcohol a ketone is produced and no further oxidation occurs. The colour of the reaction mixture changes from orange to green.

> You need to learn the reaction conditions for the oxidation of alcohols.

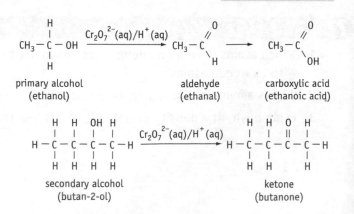

Tertiary alcohols do not undergo oxidation with acidified potassium dichromate(VI) The colour of the reaction mixture does not change, but remains orange.

Dehydration of alcohols

Alcohols can lose a molecule of water to produce an alkene. This is known as **dehydration** and is an example of an elimination reaction.

Typical reaction conditions would be using an Al_2O_3 catalyst at 400°C or heating with concentrated sulphuric acid. An example is shown below:

butan-1-ol ⟶ but-1-ene + H_2O

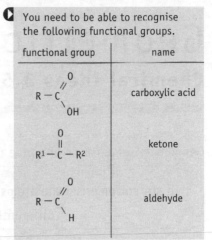

You need to be able to recognise the following functional groups.

functional group	name
R—C with =O and —OH	carboxylic acid
R^1—C—R^2 with =O	ketone
R—C with =O and —H	aldehyde

? Quick check questions

1 Draw skeletal formulae for the two oxidation products that could be obtained from the oxidation of

2 What colour change would you observe when 2-methylpropan-2-ol was heated under reflux with acidified potassium dichromate solution? Explain your answer.

3 Name the product from the dehydration of hexan-1-ol.

Geometric isomerism

Chemical Ideas 3.5

Geometric isomerism is one type of stereoisomerism.

> In stereoisomerism the atoms are bonded in the same order but are arranged differently in space in each isomer.

There are two ways of putting together the atoms of the molecule C_4H_8, but-2-ene:

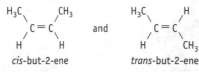

cis-but-2-ene   trans-but-2-ene

- The isomer that has the two methyl groups on the same side of the double bond is called the *cis* isomer, i.e. *cis*-but-2-ene.

- The isomer that has the two methyl groups on opposite sides of the double bond is called the *trans* isomer, i.e. *trans*-but-2-ene.

The reason these two isomers exist is that to turn one form into the other you need to break one of the bonds in the carbon–carbon double bond. There is not enough energy at room temperature to enable this to occur. As a result interconversion of the two geometric isomers does not occur.

When ethyne (C_2H_2) undergoes polymerisation it can form either *cis*-poly(ethyne) or *trans*-poly(ethyne).

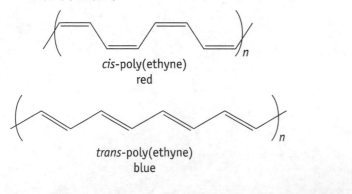

cis-poly(ethyne)
red

trans-poly(ethyne)
blue

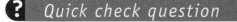

Quick check question

1 Draw and label the two geometric isomers of 1,2-dibromoethene ($C_2H_2Br_2$).

Polymers: historical developments

Chemical Storylines PR

Many of the polymers we know today were discovered by accident. You need to know some examples.

Discovery of poly(ethene)

In 1933 Gibson and Fawcett, carried out a reaction between ethene and benzaldehyde at a pressure of 2000 atm. The apparatus leaked and they obtained, by accident, a white waxy substance which had an empirical formula of CH_2.

They were then joined by Michael Perrin, who showed that exactly the right amount of oxygen was needed. He also showed that benzaldehyde could be left out and the polymerisation of ethene would still occur.

Different kinds of poly(ethene)

The polymer produced by Fawcett and Gibson was **low density poly(ethene) (ldpe)**. The chains are not well organised and so the density and the strength are lowered.

Karl Ziegler developed **high-density poly(ethene) (hdpe)**, by passing ethene through a tiny amount of $TiCl_4$ and $(C_2H_5)_3Al$ in a liquid alkane. This discovery was first prompted by the accidental formation of hdpe in apparatus that had traces of nickel left over as impurities.

Ziegler–Natta catalysts

Giulio Natta used Ziegler's catalyst to polymerise propene in 1954. His reaction mixture contained a crystalline form and an amorphous form, which he was able to separate.

- The crystalline form contained polymer chains with the methyl groups all in the same orientation. Natta called these **isotactic** forms (see page 63).

- The amorphous polymer had chains where the methyl groups were arranged at random. These he called **atactic** (also see page 63).

Natta developed new catalysts where the polymer molecule grew outward from the catalyst surface. These are called **Ziegler–Natta** catalysts.

Conducting and light-emitting polymers

In 1971, Hideki Shirakawa and Sakuji Ikeda directed a stream of ethyne, C_2H_2, on to the surface of a solution of a Ziegler–Natta catalyst at –78°C. A red film was formed.

If the experiment was repeated at 100°C the film was coloured blue. This was poly(ethyne), which can be made to conduct electricity. A student, by mistake, used 1000 times too much catalyst and produced a silvery, metallic looking film. This was the *trans* form of poly(ethyne).

Experimental techniques

The important techniques for the AS course are:

- measuring enthalpy changes (see 'Enthalpy and entropy' on page 20 and Activity DF1.2)

- making up a standard solution (see Activity EL2.1)

- using a handheld spectroscope (see Activity EL4.3)

- making thermochemical measurements (see Activity DF1.2)

- carrying out a titration (see Activity EL2.1)

- vacuum filtration (see Activity M2.3)

- fractional distillation (see Activity A4.2).

Diagrams of experimental equipment you have used

The two diagrams which are most likely to come up are the ones for **vacuum filtration** and **distillation**.

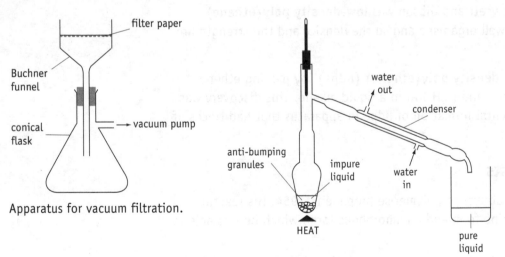

Apparatus for vacuum filtration.

Apparatus for fractional distillation.

Diagrams of experimental set-ups that you have to design

Don't panic! The information in the question stem will probably help you. Don't invent equipment, but use normal laboratory equipment. Think about what you are being asked to do, and then try to think of something which you have done that is similar.

Rules for diagrams:
- draw in cross-section, not 3D
- check the joints look airtight
- make sure anything which is supposed to move through the apparatus can do so
- use an arrow labelled 'heat' for a Bunsen burner
- always label clearly.

Vacuum filtration is a fast method of separating a solid from a filtrate.

Distillation is used to separate liquids with different boiling points. The thermometer records the boiling point of the distillate.

Example: Draw a labelled diagram of the apparatus you could use to heat zinc carbonate and measure the volume of carbon dioxide evolved.

	would gain full marks	would not gain full marks
tube connected with no leaks	✓	✗
gas syringe (or collection over water)	✓	✓
heat source labelled	✓	✗
calibration of collection vessel shown	✓	✗

Example: Bromine can be obtained in the laboratory by the reaction between aqueous potassium bromate(V) and sulphur dioxide. Draw a labelled diagram of a simple apparatus you could use safely to carry out this reaction in the laboratory. You are provided with a sulphur dioxide supply from a gas cylinder.

> Remember to show the calibration if you are measuring volume. Check the gas can't leak.

Your diagram needs to have:

- a tube labelled 'sulphur dioxide' with a bubbler through a labelled solution in recognisable laboratory apparatus
- some indication of safety, e.g. 'do in a fume cupboard' or a tube shown taking gas away to safe disposal.

Describing experimental techniques

Sometimes a question asks you to describe an experimental technique you have met.

Example: The amount of iron in an iron tablet can be measured by titrating a solution of the tablet in sulphuric acid with potassium manganate(VII) solution. Describe the main steps involved in such a titration (4 marks).

> Note that you need four points for 4 marks.

- Pipette out a known volume of tablet solution.
- Titrate with potassium manganate(VII) solution in a burette ...
- ... until the pink or purple colour stays.
- Repeat the titration.

Exam hints and tips

Here are some general points that apply to both Module 2850 (Chemistry for Life) and Module 2848 (Chemistry of Natural Resources) written examination papers.

1 **All the questions are structured.** This means that you are given a 'stem' of information which provides the context (the Storyline) for the question. This is followed by a series of part-questions. It can be quite helpful to underline key pieces of information as you read through the stem.

2 **The questions are designed so that you work through them in stages.** Answers to the part-questions are often linked together and are linked to the information you are given. Work through the questions in order – don't cherry pick.

3 **Part questions are linked by the context** – not by chemical topic. You will need to dip into several different parts of your knowledge in one question.

4 **Contexts will be a mixture of familiar ones from the Storylines and unfamiliar ones.** Don't panic if the context is unfamiliar – the chemistry you are being asked will be familiar.

5 **The Data Sheet provides additional information not given in the question.** Make sure you are familiar with what is on the Data Sheet. Remember to use it.

6 **All questions are compulsory – you have no choice.** Try to answer every question. It is better to make a sensible guess, which could score you one mark, than to put nothing at all. You cannot get negative marks. You have nothing to lose by guessing – so use your chemical common sense!

7 **Examiners use an agreed mark scheme.** The marks are given for very specific points, so you must be precise and use language accurately.

8 **Some questions will require a knowledge of applications of chemistry and the work of chemists.** This means that it is a good idea to have read and made notes on the Storylines, particularly those sections listed in the 'Check your Notes' activities at the end of each unit.

9 **The first question is designed to be quite straightforward**, to help settle you into the exam. So, you are advised to do the questions in the order they are set.

Some tips:

- Answer what is asked.
- Give a sign with oxidation states: a '+' or '−' before the number (e.g. +3, −5).
- Enthalpy changes must have a sign as well as a unit.
- Give state symbols in chemical equations only when requested. They are often asked for in reactions where there is a change of state.
- Use chemical language correctly (e.g. make sure you know the difference between atom, molecule and ion, between chlorine and chloride, and between hydroxide and hydroxyl).

Types of examination questions

'Explain' questions

These are not asking for formal definitions, but for you to describe in your own words what is meant by a chemical term or concept, or for you to explain an observation. These part-questions can often require some extended writing. Some will be used to assess the quality of your written communication – you will always be told if this is the case.

> To get your written communication mark, you must write in full sentences (no bullet points). Make sure you are presenting a logical account using correct scientific terms.

Economic, environmental and social questions

These questions can sometimes sound vague, but require specific answers. You must be careful to give an answer in terms of what you have learnt in your AS chemistry course. Use the information you have been given in the question.

Writing chemical formulae questions

You will be asked to draw chemical formulae for organic molecules.

- A **full structural formula** means that you must show every bond in the molecule. A common error is to miss some of the hydrogen atoms. Alcohol groups should be written –O–H, not –OH. You do not need to draw out all the bonds in benzene though – use the usual hexagon with a circle inside.

> Try drawing a full structural formula of pentanoic acid.

- If you are asked simply to draw the **structure** of a molecule, be guided by the style of similar molecules in the question. You do not need to draw out every bond, but the functional groups should be clear and the structure unambiguous.

- A **skeletal formula** shows the carbon skeleton as angled lines in a zig-zag – carbon atoms are assumed to be at the ends of each line. Functional groups then stand out very clearly. Hydrogen atoms are shown on functional groups (but not those directly on the carbon skeleton).

> Try drawing the skeletal formula of propanone.

- Dot-cross formulae should show lone (non-bonding) pairs of electrons as well as bonding electrons.

> Try drawing a dot-cross formula for N_2H_4.

Calculation questions

Always show your working. Examiners use consequential marking where possible to give credit for correct working even if your final answer is wrong, so make sure you include everything. Even simple things like calculating numbers of moles can get you a mark.

Here are some other calculation do's and don'ts:
- write your final answer on the answer line
- always give units (unless already present) and a sign if appropriate (a sign is <u>always</u> needed for enthalpy changes)
- give the answer to the same number of significant figures as the data in the question (usually 2 or 3 sf)
- don't give the answer alone
- always think about whether your answer is sensible (e.g. ΔH_c values are always negative). If you think the answer is not sensible and you can't see how to change it, it is worth commenting on this to the examiner.

Practice exam questions

Module 2850

1 Mass spectrometry can be used to identify ivory which has been traded illegally. The ratio of ^{12}C to ^{13}C differs in the ivory from different regions, enabling the origin of a sample to be traced.

    (a) **(i)** Complete the table to show the numbers of particles present in atoms of ^{12}C and ^{13}C.   **[4]**

name of particle	^{12}C atom	^{13}C atom
protons		
neutrons		
electrons		

       (ii) Write down the atomic number and the mass number of ^{12}**C**.   **[2]**

    (b) What term is used to describe two forms of the same element, such as ^{12}C and ^{13}C?   **[1]**

> 'Iso' means 'the same', as in 'isomer' and 'isotope'.

    (c) A diagram of a mass spectrometer is shown below.

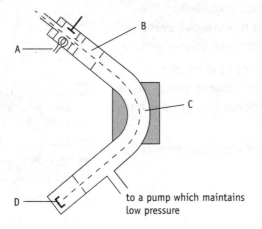

to a pump which maintains low pressure

       Particles are ionised at **A** and then accelerated at **B**.

       Explain how they are then separated at **C** so that ions of different masses can be detected at **D**.   **[3]**

> What is altered so that each ion in turn hits the detector?

    (d) A sample of ivory contains 98.89% of ^{12}C and 1.11% of ^{13}C. Calculate the relative atomatic mass of carbon in the sample to two decimal places.   **[3]**

> Have you read the question? Check your decimal places. Do you need units?

    (e) Another form of carbon is ^{15}C. This has a radioactive nucleus that decays, giving β-particles and γ-rays.

       (i) Complete the nuclear equation for the decay of ^{15}C to give a β-particle.
$$^{15}_{6}C \rightarrow$$
           [3]

> Use the Periodic Table.

       (ii) A sample of a solid carbon compound containing ^{15}C is to be stored safely. Explain how ^{15}C is hazardous and suggest how it should be stored, giving reasons.   **[4]**

> Go back and read the stem for part **e**. Two types of radiation are emitted.

From OCR June 2003, Module 2850   **[Total: 20]**

2 The tendency of a fuel to auto-ignite in a petrol engine is measured by its octane rating. The hydrocarbon 2,2,4-trimethylpentane has a low tendency to auto-ignite and is given an octane rating of 100. Heptane has a high tendency to auto-ignite and is given an octane rating of zero.

(a) **(i)** Name the homologous series to which both 2,2,4-trimethylpentane and heptane belong. [1]

(ii) Draw the structural formula of **heptane**, which has seven carbon atoms in its molecule. [1]

> 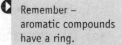 Structural formulae show bonds.

(iii) Draw the **skeletal** formula of 2,2,4-trimethylpentane. [2]

> Check the number of carbons in the pentane chain.

(b) Draw the structural formula of an isomer of **heptane** and give its systematic name. [2]

(c) To measure the octane rating of a fuel, the mixture of 2,2,4-trimethylpentane and heptane is found which auto-ignites under the same conditions as the fuel. The percentage of 2,2,4-trimethylpentane in this mixture gives the octane rating.

'Four-star petrol' has an octane rating of 97. Give the composition of a mixture of hydrocarbons that would have the same auto-ignition properties. [1]

(d) Three compounds that have high octane ratings are shown below.

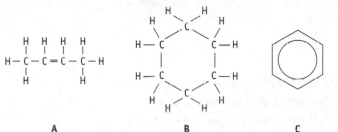

| A | B | C |

(i) To which homologous series do compounds **A** and **B** each belong? [2]

(ii) State, with a reason, which of the three compounds is/are aromatic. [2]

> Remember – aromatic compounds have a ring.

(e) Compound A can be made by cracking heptane.

Draw a labelled diagram of the apparatus you would use to crack a sample of liquid heptane and collect the gaseous products. [3]

>  Don't panic! Remember to label the diagram and check for leaks. You probably did this experiment with paraffin at GCSE.

(f) **(i)** Complete the dot-cross diagram for **compound A**.

full structural formula for compound A   dot-cross diagram for compound A   [2]

(ii) Give the values of the bond angles **P** and **Q** in the molecule of **compound A**. [2]

> The angles won't be 90° as in this diagram.

(g) (i) Balance the chemical equation below for the complete combustion of **compound A**. [2]

Double check your balancing.

$$C_4H_8 + ...O_2 \rightarrow ...CO_2 + ...H_2O$$

(ii) Use the balanced equation and the data given below to calculate a value for the enthalpy change of combustion of **compound A**.

Draw the molecules and make sure you count ALL the bonds.

Bond	Bond enthalpy /kJ mol^{-1}
C—C	+347
C—H	+413
O=O	+498
C=O	+805
O—H	+464
C=C	+612

[4]

(iii) Give **two** reasons why the value from (ii), correctly calculated, will not be the same as a value from a Data Book. [2]

If you are stuck, look back at the section on bond enthalpies (page 22).

(iv) The C—C bond is longer than the C=C bond. What evidence is there for this from the table in (ii)? [1]

From OCR June 2003, Module 2850 [Total: 27]

Module 2848

3 This month the Canadian Space Agency should be launching a satellite mission to monitor the chemical processes that control the distribution of ozone in the atmosphere. It will focus on the Arctic winter stratosphere (upper atmosphere). The satellite uses the Sun as a source of infrared, visible and ultraviolet radiation. It will measure how much of each wavelength in these ranges is absorbed by the stratosphere.

(a) (i) Describe what can happen to molecules when they absorb infrared radiation and ultraviolet radiation. [6]

Write in full sentences. Use scientific words correctly and use correct spelling, punctuation and grammar to obtain this mark.

(*In this question, 1 mark is available for the quality of written communication.*)

(ii) Previous experiments have shown that in winter the concentration of ozone in the stratosphere falls. The infrared spectrum of ozone contains a peak at 2105 cm^{-1}.

The diagram below shows how this peak looks in the summer.
On the diagram below, draw a peak to represent how this part of the infrared spectrum would appear in winter. [2]

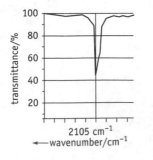

(b) The Canadian scientists plan to monitor the concentrations of about 30 different gases, such as CF_2Cl_2, in the atmosphere. Some of these gases may break down to produce radicals. Radicals can lead to the thinning of the ozone layer.

(i) What is a *radical*? [1] ▶ A definition is needed here.

(ii) CF_2Cl_2 can break down in the stratosphere to form radicals. Describe how this occurs. [3] ▶ Is this homolytic fission or heterolytic fission?

(iii) When CF_2Cl_2 breaks down in the stratosphere, chlorine radicals are produced but **not** fluorine radicals. What is the reason for this difference? [2] ▶ Think about relative bond enthalpies within the molecule.

(iv) Chlorine radicals can have a catalytic effect on the destruction of ozone. Complete the **two** equations given below to show this. Underneath, write the **overall** equation. [3] ▶ Add the two equations together. Then cancel things that appear on both sides of the overall equation.

$$Cl + O_3 \rightarrow$$
$$\rightarrow O_2 + Cl$$

overall equation $\rightarrow$

(c) The molecule Cl_2O is also present in the stratosphere.

(i) Draw a diagram to show the shape of the molecule. [2] ▶ Remember to think about bond angles.

(ii) Use the electronegativity data given below to deduce the polarity of the Cl–O bond.

Now use the shape you have drawn for the Cl_2O molecule. Decide whether Cl_2O is a polar molecule or not. Give your reasoning. [Electronegativity O, 3.4; Cl, 3.2] [3]

From OCR June 2002, Module 2851 (now included in 2848) [Total: 22]

---

4 A 'superglue' works by the polymerisation of the alkene **A**.

$$\underset{H}{\overset{H}{\diagdown}}C = C\underset{COOC_2H_5}{\overset{CN}{\diagup}}$$ **alkene A**

Alkene **A** polymerises very easily when in contact with a nucleophile.

(a) (i) Name the **type** of polymerisation which occurs when alkene **A** polymerises. [1]

(ii) Draw the structure of the repeating unit of the polymer which would be formed. [1] ▶ Show one repeating unit in brackets.

(iii) Name the functional group present in $-COOC_2H_5$. [1] ▶ Give the name, not the structure here.

(b) The polymerisation of alkene **A** is catalysed by amines in human skin because they are nucleophiles.

(i) Give the structure of the amine functional group. [1]

(ii) Suggest which feature of the amine group makes it a nucleophile. [1]

(c) Nucleophilic attack occurs at the left-hand carbon atom of alkene **A**. This atom is made slightly positively charged because of withdrawal of electrons in the double bond towards the other carbon atom. This occurs because the carbon–nitrogen and the carbon–oxygen bonds are polar.

(i) Explain what you understand by the term *polar* bond. [2]

> Think about electronegativities here.

(ii) Explain how the bond polarity is related to the electronegativity differences of the carbon, nitrogen and oxygen atoms. [3]

(iii) The polymerisation of alkene **A** takes place by a nucleophilic mechanism. What type of mechanism is involved in the polymerisation of ethene to form low-density poly(ethene)? [1]

(d) (i) State the **types** of intermolecular forces you would expect between chains of the polymer of alkene **A**. [1]

> There are two different forces to mention.

(ii) The –CN group (present in alkene **A**) is replaced in alkene **B** by a –CH$_3$ group which has a **similar size** to a –CN group.

alkene A   alkene B

Which would you expect to be more flexible, the polymer of alkene **A** or the polymer of alkene **B**? Explain your answer. [3]

> Think about what causes flexibility, then decide which polymer would be most flexible.

(e) Does alkene **B** show *cis–trans* isomerism? Explain your answer. [1]

From OCR June 2000, Module 5682 (now included in 2848) [Total: 16]

5 The element zirconium, Zr, and its compounds are becoming increasingly important as constituents in a wide range of high-tech materials. The main ore of the element is called 'zircon sand'.

(a) The atomic number of zirconium is 40. Which block in the Periodic Table is zirconium in? [1]

> Use your Periodic Table.

(b) 'Zircon sand', ZrSiO$_4$, is used to make zirconium(IV) oxide.

(i) The equation for the first stage in this process is given below.

Balance this equation by writing numbers on the dotted lines.

...NaOH + ZrSiO$_4$ → Na$_2$SiO$_3$ + Na$_2$ZrO$_3$ + ...H$_2$O [2]

(ii) In the next stage of the process, water is added. The result is a precipitate of zirconium(IV) oxide in an aqueous solution. Suggest a method of separating the solid from the aqueous solution. [1]

> Think through laboratory techniques you have used.

(iii) Calculate the maximum mass of zirconium(IV) oxide, ZrO$_2$, which is produced from 1000 kg of pure ZrSiO$_4$. Assume 1 mole of ZrSiO$_4$ gives 1 mole of ZrO$_2$.
[A_r: 0,16; Si,28; Zr,91] [2]

> Show your working out clearly.

(c) The aqueous solution from (b) (ii) contains sodium silicate, Na_2SiO_3. Silicon is in Group 4 of the Periodic Table and in some ways its chemistry resembles that of carbon.

(i) Silicon has a similar outer electron structure to carbon. In the boxes below, show the s and p electrons of silicon's electron structure. [2]

(ii) Both carbon and silicon from covalent dioxides, CO_2 and SiO_2. These have very different structures and properties. Describe **one** way in which CO_2 and SiO_2 differ in their **properties**. [2]

(iii) How does the **structure** of SiO_2 differ from that of CO_2? [2]

(iv) Use your answer to (iii) to help you **explain** why CO_2 and SiO_2 differ in their properties. [2]

(d) One of the most important uses of zirconium(IV) oxide is as a catalyst. It is used in catalytic converters for cars because it can withstand high temperatures.

(i) The diagram below shows an enthalpy profile for an uncatalysed reaction.

- draw the enthalpy profile for the same reaction in the presence of a catalyst
- mark clearly on the diagram the **activation enthalpy** of the **catalysed** reaction
- mark clearly on the diagram the **activation enthalpy** of the **uncatalysed** reaction [5]

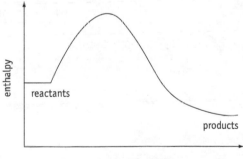

progress of reaction

(ii) Use the diagram and collision theory to explain why reactions go faster in the presence of a catalyst. [3]

From OCR January 2003, Module 2851 (now included 2848) [Total: 22]

> Remember parallel spins!

> Give a property of each and then compare them.

> Describe each compound.

> Give reasons for your answer to part (iii).

> Take care to label your answers on the graph clearly.

> Use the diagram. Think of AS answers, not GCSE.

Mark schemes – Module 2850

Question	Expected answers	Marks
1 a i	$\quad\quad\quad\quad\quad\quad\quad\quad$ C-12 $\quad\quad$ C-13 protons $\quad\quad\quad\quad\quad\quad$ 6 $\quad\quad\quad\quad$ 6 $\quad\quad\quad$ (1) (both correct) neutrons $\quad\quad\quad\quad\quad$ 6 $\quad\quad\quad\quad$ 7 $\quad\quad\quad$ (2) (1 each) electrons $\quad\quad\quad\quad\quad$ 6 $\quad\quad\quad\quad$ 6 $\quad\quad\quad$ (1) (both correct)	4
1 a ii	atomic number 6; mass number 12.	2
1 b	isotope(s) (A few candidates wrongly said isomers.)	1
1 c	*3 from 4:* • (electro)magnet(ic field) (magnet is the key word needed) • deflects/attracts/repels/bends/anything implying change of direction $\quad$ ions • according to their mass/weight(/charge ratio)/heavy light; NOT large/small • varying the magnetic <u>field</u> brings the particles on to **D**. $\quad$ (marks can be scored for labels on diagram)	3
1 d	$(98.89 \times 12) + (1.11 \times 13) (= 1201.11)$ (1) $\div 100$ $= 12.01$ (2) No units. If not 2 dp, then max of 2 marks. Just 12.01 scores 3.	3
1 e i	$^{15}_{6}C \rightarrow {}^{0}_{-1}e + {}^{15}_{7}N$ e (1), N (1), all extra detail (1)	3
1 e ii	*Any four from:* β-particles/γ-rays are ionising/oxidising (1) destroy DNA/cause mutations/cancer/damage cells/skin/tissue (1) NOT people/us etc NOT harmful β can penetrate skin/stopped by (any) metal (foil)/stone/concrete (1) γ great penetration/(only) stopped/absorbed by lead (1) (a lot of candidates forgot about the γ-radiation) <u>lead</u> container needed to protect (1)	4
	Total	20

Question	Expected answers	Marks
2 a i	alkane(s)	1
2 a ii	CH_3–CH_2–CH_2–CH_2–CH_2–CH_2–CH_3 etc (or full structural)	1
2 a iii	Five carbon backbone (1) Correct branches NOT CH_3(1) Max (1) for "blobs", allow dots.	2
2 b	structure (as a ii) (1); name (*must match structure of carbon backbone*) *ignore commas, dashes, gaps*) (1)	2
2 c	97% 2,2,4-trimethylpentane and 3% heptane	1
2 d i	alkene(s) (1); <u>cyclo</u>alkane(s) (1)	2
2 d ii	C only (1); has a <u>benzene</u> ring/delocalised electrons/is benzene/arene (1) *depends on first being scored*	2
2 e	 • heating tube or boiler and tube with solid, connected without leaks to … (1) • collection tube (or gas syringe) (1) *mark separately* • labels, minimum heptane, catalyst (ignore qualification)/alumina/pot, catalyst heated (1) (can award label mark if <u>catalyst</u> in heptane)	3
2 f i	 double bond (any combination of two dots and two crosses or other symbols (1); single bonds (1)	2
2 f ii	P = 109 (±2) (1); Q = 120 (±3) (1) (degree sign not needed)	2
2 g i	$C_4H_8 + 6O_2 \rightarrow 4CO_2 + 4H_2O$ products (1); reactants (1) *any balanced equation*	2

2 g ii	Broken 2 × C–C 694 Made 8 × C=O 6440
	            8 × C–H 3304 8 × O–H 3712
	            1 × C=C 612
	            6 × O=O 2988
	Total 7598 (1) 10152 (1)
	Broken – made (1) = 7598 – 10152 = –2554 kJ mol⁻¹ (1) *including sign and unit*
	NOT/mol⁻¹ ALLOW –2550 kJ mol⁻¹
		4

2 g iii	Average/approx <u>bond enthalpies</u> used/vary between molecules (1); Standard <u>states</u> not used/reference to $H_2O(l)$ (1)	2
2 g iv	(C–C has a) smaller bond enthalpy/weaker bond/less energy to break. Must be a *comparison*.	1
	Total	27

Mark schemes – Module 2848

Question	Expected answers	Marks
3 a i	*For infrared*: bond (1); vibrates (1); to higher vibrational level (1); *For ultraviolet*: **three** *points max* electron(s) (1); excited (1) *or* move to higher energy level (1); bonds are broken/form radicals (1); molecules are ionised (1). *At least two readable and clear sentences with no more than one spelling, punctuation or grammatical error* (1)	6
3 a ii	Both peaks at same frequency Winter peak smaller than summer peak	2
3 b i	Atom or molecule having at least one unpaired electron	1
3 b ii	*Any 3 points from 4:* Absorption/hv (1); of ultraviolet/high frequency radiation (1); causes <u>bonds</u> to break (1); homolytically (1)	3
3 b iii	Bond enthalpy (1); of C–Cl is smaller than C–F (1)	2
3 b iv	$Cl + O_3 \rightarrow ClO + O_2$ (1) $ClO + O \rightarrow O_2 + Cl$ *or* $ClO + O_3 \rightarrow 2O_2 + Cl$ (1) *Overall*: $O + O_3 \rightarrow 2O_2$ *or* $2O_3 \rightarrow 3O_2$ (1)	3
3 c i	Non-linear *or* bent triatomic *molecule drawn* (1); *order of atoms is* ClOCl (1)	2
3 c ii	Compares electronegativity (1); $^{\delta+}Cl - O^{\delta-}$ (1) Dipoles of bonds do not cancel therefore molecule polar (1)	3
	Total	22

Question	Expected answers	Marks
4 a i	addition	1
4 a ii	$\left(\begin{array}{cc} H & CN \\ | & | \\ -C - C- \\ | & | \\ H & COOC_2H_5 \end{array} \right)_n$	1
4 a iii	ester	1
4 b i	$-NH_2$	1
4 b ii	lone pair of electrons	1
4 c i	electrons unequally shared (1); leading to a partial separation of charge (1)	2
4 c ii	N <u>and</u> O are more electronegative than C (1) thus they pull electrons from C *or* electronegativity is the power of an atom in a molecule/covalent bond to attract electrons to itself (only one of O and N need be mentioned here.) so carbon positively charged *or* oxygen/nitrogen negatively charged (1) *or* the greater the electronegativity difference the more polar the bond (1)	3

Question	Expected answers	Marks
4 c iii	radical	1
4 d i	permanent dipole–permanent dipole instantaneous dipole-induced dipole	1
4 d ii	polymer of B because <u>weaker</u> intermolecular forces result between chains (not 'less' or 'fewer') (1); since CH_3 is non-polar (1) the chains can therefore move past each other more easily (1)	3
4 e	No, two hydrogens/identical groups on one C/one side of bond consideration of possible isomers	1
	Total	16

Question	Expected answers	Marks
5 a	d block	1
5 b i	$4NaOH + ZrSiO_4 \rightarrow Na_2SiO_3 + Na_2ZrO_3 + 2H_2O$	2
5 b ii	Filtration	1
5 b iii	$M_r (ZrSiO_4) = 183$ & $M_r (ZrO_2) = 123$ (1); Mass of $ZrO_2 = 1000 \times 123/183 = 672$ kg (1)	2
5 c i	3s 3p [↑↓] [↑ ↑] *1 mark for electrons;* *1 mark for detail correct.*	2
5 c ii	SiO_2 is a solid (1) CO_2 is a gas (1) (or expressed in terms of relative melting/boiling points) or SiO_2 is insoluble in water (1) CO_2 is soluble (1)	2
5 c iii	SiO_2 has giant (covalent) structure (1); CO_2 has a molecular structure (1)	2
5 c iv	intermolecular bonds/forces between molecules are easier to break (1); than (chemical) bonds in a giant structure (1)	2
5 d i		5
5 d ii	Catalysed reaction has lower activation enthalpy (1); more collisions will have the necessary activation enthalpy (1); more collisions will lead to products (1)	3
	Total	22

Answers to quick check questions

Module 2850

The Elements of Life (EL)

Amount of substance

Page 2

1 (a) 40 (b) 142 (c) 323

2

	Beryllium in 100 g	Carbon in 100 g	Oxygen in 100 g
Step 1 Masses	12.9	17.3	69.8
Step 2 Divide by A_r...	$\dfrac{12.9}{9}$	$\dfrac{17.3}{12}$	$\dfrac{69.8}{16}$
...to give amount in moles	1.43	1.44	4.36
Step 3 Divide by smallest number...	$\dfrac{1.43}{1.43}$	$\dfrac{1.44}{1.43}$	$\dfrac{4.36}{1.43}$
...to give simplest ratio of moles	1 mol	1 mol	3 mol

Formula is $BeCO_3$.

3 percentage by mass =

$$\frac{\text{mass of element in 1 mole of the compound}}{M_r} \times 100$$

$$= \frac{207}{323} \times 100 = 64.1\%$$

Periodicity and the Periodic Table

Page 5

1 The melting point increases then decreases as you go across the 3rd Period.

2 880 kJ mol^{-1} (± 20)

3 The nuclear charge is greater in calcium than in potassium so there is a greater attraction between the nucleus and outer shell electrons. This means that more energy is needed to remove the electron in calcium.

The outer electron in potassium is further away from the nucleus than in sodium. There is less attraction between the nucleus and the outer shell electron in potassium so less energy is needed to remove it.

4 $Ca^+(g) \rightarrow Ca^{2+}(g) + e^-$

A simple model of the atom

Page 7

1 (a) 1p, 1e$^-$, 2n (b) 20p, 20e$^-$, 27n (c) 52p, 52e$^-$, 70n (d) 95p, 95e$^-$, 146n (e) 8p, 10e$^-$, 8n

2 $\dfrac{(24 \times 70) + (25 \times 19) + (26 \times 11)}{100} = 24.4$

3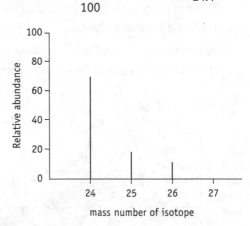

Nuclear reactions

Page 9

1 (a) $^{47}_{21}Sc$ (b) $^{222}_{86}Rn$ (c) $^{237}_{93}Np$ (d) $^{131}_{54}Xe$

2 No, the alpha particles will be stopped/absorbed by the watch.

3 ½ will be left after 6.5 days, ¼ will be left after 13 days, ⅛ will be left after 19.5 days and ¹⁄₁₆ will be left after 26 days.

4 Geiger counter

Light and electrons

Page 11

1 (a) A black background with coloured lines on it.
  (b) There are no electrons in a hydrogen nucleus.

2 At least three horizontal lines. Lines should become closer as the energy increases. Two different vertical lines pointing upwards between levels.

3 Electrons are in energy levels. The electrons absorb energy and move to a higher energy level. The frequency of light absorbed is related to the difference in energy levels by $\Delta E = hv$.

4 C is 2.4, Si is 2.8.4

5 Strontium and calcium both have two electrons in their outer shell.

6 The outer electron in potassium is further from the nucleus than the outer electron in sodium. The outer electron is less firmly held by electrostatic attraction in potassium and so is more easily lost.

Chemical bonding and shapes of molecules

Page 14

1 Covalent, ionic, covalent, metallic, metallic.

2

3 $a = 109°$, $b = 109°$, $c = 109°$, $d = 180°$, $e = 120°$, $f = 120°$

Group 2 and balancing equations

Page 15

1 $Sr(s) + 2H_2O(l) \rightarrow Sr(OH)_2(aq) + H_2(g)$

2 (a) $CaCl_2$ (b) $Sr(OH)_2$

3 Use universal indicator or another named indicator. pH > 7 or a colour change for the named indicator.

4 $CaCO_3(s) \rightarrow CaO(s) + CO_2(g)$

5 (a) $Ba(OH)_2$ (b) $MgCO_3$

Developing Fuels (DF)

Calculations from equations

Page 19

1

Step 1 Underline the substances whose:	4Fe	+ 3O$_2$	$\rightarrow$ 2Fe$_2$O$_3$
• mass you are given			
• mass you want to find			
Step 2 Indicate moles involved	4 moles		2 moles
Step 3 Calculate the masses	224 g		320 g
Step 4 Convert to the mass given in the question	$\frac{224}{224} \times 11.2 = 11.2$ g		
Step 5 Convert the other mass by the same amount			$\frac{320}{224} \times 11.2 = 16.0$ g
Step 6 Write down the answer			16 g of Fe$_2$O$_3$ is produced

2

Step 1 Underline the substances whose:	6CO	+ 13H$_2$	$\rightarrow$ C$_6$H$_{14}$ + 6H$_2$O
• volume you are given			
• volume you want to find			
Step 2 Indicate moles involved	6 moles	13 moles	
Step 3 Calculate the volumes	144 dm^3	312 dm^3	
Step 4 Convert to the volume given in the question	$\frac{144}{144} \times 6 = 6.0$ dm^3		
Step 5 Convert the other volume by the same amount		$\frac{312}{144} \times 6 = 13$ dm^3	
Step 6 Write down the answer		13 dm^3 of H$_2$ is needed	

3

Step 1 Underline the substances whose:		$2NaN_3$	$\rightarrow 2Na$	$+ 3N_3$

Step 1 Underline the substances whose:
- mass or volume you are given
- mass or volume you want to find

	$\underline{2NaN_3}$	$\rightarrow 2Na$	$+ \underline{3N_3}$
Step 2 Indicate moles involved	2 moles		3 moles
Step 3 Calculate the mass or volume	130 g		72 dm³
Step 4 Convert to the mass or volume given in the question	$\frac{130}{130} \times 0.65 = 0.65$ g		
Step 5 Convert the other mass or volume by the same amount			$\frac{72}{130} \times 0.65 = 0.36$ dm³
Step 6 Write down the answer			0.36 dm³ of N_2 is produced

4

	$\underline{C_8H_{18}}$	$+ 12.5O_2$	$\rightarrow 8CO_2$	$+ 9H_2O$	$\Delta H = \underline{-5470}$ kJ mol⁻¹

Step 1 Underline
- the mass you are given
- the enthalpy change you are given

		$\Delta H = \underline{-5470}$ kJ mol⁻¹
Step 2 Indicate moles involved	1 mole	
Step 3 Calculate the mass	114 g	
Step 4 Convert to the mass or volume given in the question	$\frac{114}{114} \times 5.7 = 5.7$ g	
Step 5 Convert the enthalpy change by the same amount		$\frac{-5470}{114} \times 5.7 = -273.5$ kJ mol⁻¹
Step 6 Write down the answer		273.5 kJ of energy is released

Enthalpy and entropy

Page 20

1 Draw and label the following: propanol in the burner, under a container of water, thermometer in the water and a draught shield around the burner.

2 Liquids have a lower entropy than gases <u>or</u> there are fewer ways of arranging liquid particles <u>or</u> there are fewer particles formed.

Hess's law

Page 21

1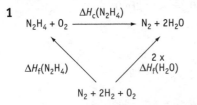

2 $\Delta H_c(CO) = \Delta H_f(\text{products}) - \Delta H_f(\text{reactants})$

$= -393.5 - (-110.5) = -283$ kJ mol⁻¹

3 $\Delta H_c(N_2H_4) = \Delta H_f(\text{products}) - \Delta H_f(\text{reactants})$

$= 2 \times (-286) - (+51) = -623$ kJ mol⁻¹

Bond enthalpies

Page 22

1
Bonds broken		Bonds made	
$1 \times$ H–H (436)	= 436	$2 \times$ O–H (464)	= 928
$0.5 \times$ O=O (498)	= 249	Total	= –928
Total	= +685		

$\Delta H = +685 - 928 = -243$ kJ mol⁻¹

2
Bonds broken		Bonds made	
$5 \times$ C–C (347)	= 1735	$14 \times$ C=O (805)	= 11 270
$1 \times$ C=C (612)	= 612	$14 \times$ O–H (464)	= 6496
$14 \times$ C–H (413)	= 5782		
$10.5 \times$ O=O (498)	= 5229		
Total	= +13 358	Total	= –17 766

$\Delta H = +13\ 358 - 17\ 766 = -4408$ kJ mol⁻¹

Alcohols and ethers

Page 23

1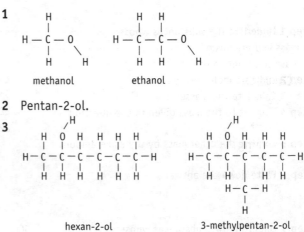

methanol   ethanol

2 Pentan-2-ol.

3

hexan-2-ol

3-methylpentan-2-ol

4 Ethoxyethane is an oxygenate so produces less carbon monoxide on burning than the corresponding alkane <u>or</u> it has a high octane number so little tendency to auto-ignite.

Alkanes and other hydrocarbons

Page 25

1 Octane

2

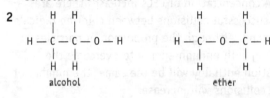

3-methylhexane   2,2,4-trimethylheptane

3   $CH_4 + 2O_2 \rightarrow CO_2 + 2H_2O$

    $C_2H_6 + 3.5O_2 \rightarrow 2CO_2 + 3H_2O$ <u>or</u>

    $2C_2H_6 + 7O_2 \rightarrow 4CO_2 + 6H_2O$

4   They all have the formula C_6H_{12}.

5

This is aliphatic because it doesn't have a benzene ring.

Structural isomerism

Page 26

1

OH   OH

2-methylpropan-2-ol   2-methylpropan-1-ol

They have the same molecular formula but different structural formulae.

2

H H   H   H

| | | |

H — C — C — O — H   H — C — O — C — H

| | | |

H H   H   H

alcohol   ether

Auto-ignition and octane numbers

Page 27

1   hexane < cyclohexane < benzene

          increasing octane number

2   Dimethylpropane (as it is not an oxygenate).

Catalysts

Page 28

1   $C_8H_{18} \rightarrow C_3H_6 + C_5H_{12}$

2   Reforming (as H_2 is produced).

Pollution from a car

Page 29

1   Incomplete combustion of a fuel/petrol.

2   Contributes to formation of photochemical smog, causing respiratory problems <u>or</u> causes acid rain which erodes buildings.

3   More complete combustion <u>or</u> more reaction with air.

4   It must be liquefied and so either low temperatures or high pressures are needed.

5   energy density (kJ kg^{-1}) = enthalpy of combustion (kJ mol^{-1}) × amount of fuel in 1 kg (mol kg^{-1})

    = 243 × 500 = 121 500 kJ kg^{-1}

Module 2848

From Minerals to Elements (M)

Ions in solids and solutions

Page 33

1   Simple cubic

2   $Ba^{2+}(aq) + SO_4^{2-}(aq) \rightarrow BaSO_4(s)$

3

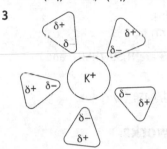

Oxidation and reduction

Page 35

1   K +1, Br –1; H +1, O –2; P +5, O –2

2   Br –1 to 0, S +6 to +4

3   $2Ca \rightarrow 2Ca^{2+} + 4e^-$ oxidation (electron loss)

    $O_2 + 4e^- \rightarrow 2O^{2-}$ reduction (electron gain)

Electronic structure: sub-shells and orbitals

Page 37

1   (a) 3 (b) s, p and d

2   $1s^2\ 2s^2\ 2p^6\ 3s^2\ 3p^4$

3   Ca

The p block: Group 7

Page 39

1   Red-brown liquid

2   $AgNO_3(aq) + KI(aq) \rightarrow AgI(s) + KNO_3(aq)$

    A yellow precipitate would form.

3   (a) $Cl_2(g) + 2Br^-(aq) \rightarrow 2Cl^-(aq) + Br_2(aq)$

    (b) The solution would turn dark brown, due to the presence of bromine.

    (c) The upper cyclohexane layer would turn red due to bromine dissolving in it.

Concentrations of solutions

Page 41

1 (a) $0.02\ dm^3$ (b) $0.02 \times 0.100 = 0.002$ moles

(c) 0.004 moles (d) $0.004 \times \dfrac{1000}{25} = 0.16$ moles

(e) $0.16\ mol\ dm^{-3}$

Acid–base reactions

Page 43

1 (a) acid = HBr, base = NH_3
(b) acid = H_3O^+, base = SO_4^{2-}

2 conjugate acid/base pairs are H_2SO_4/HSO_4^- and H_2O/OH^-

3 (a) precipitation (b) acid/base (c) redox

Molecules and networks

Page 44

1 Strong intramolecular covalent bonds form a giant 3D network, which is very difficult to break.

2 CO_2 is made up of small molecules, which can easily be pulled part. There is enough energy to do this at room temperature so it is a gas.

SiO_2 is a solid at room temperature since it consists of a covalently bonded giant network, which needs a lot of energy to break up.

The Atmosphere (A)

What happens when radiation interacts with matter?

Page 47

1 (a) 2.52×10^{-20} J (b) infrared (c) vibrational

2 (a)

$$Cl + O_3 \rightarrow ClO + O_2$$
$$ClO + O \rightarrow Cl + O_2$$

Overall $\quad O + O_3 \rightarrow 2O_2$

(b) The chlorine radical is regenerated in the reaction and can go on to catalyse further reactions.

Radiation and radicals

Page 49

1 Homolytic fission produces radicals, heterolytic fission produces ions.

2 Initiation, propagation and termination.

3 A radical has one or more unpaired electrons while a nucleophile has one or more lone pairs of electrons.

Halogenoalkanes

Page 51

1

4-Bromo-3,3-dichloro-2-iodohexane

2 Heat the halogenoalkane in a sealed tube with concentrated ammonia solution.

3 The C–I bond is weaker than the C–F bond, so it is easier to break the C–I bond.

Chemical equilibrium

Page 53

1 (a) Increase concentrations of C_2H_4 and/or H_2O
(b) lower the temperature (c) increase the pressure.

2 (a) Deepening of the red colour (b) lightening of the red colour/production of yellow colour.

Rates of reaction

Page 54

1 As the concentration of CFCs increases there are more successful collisions between chlorine radicals and ozone. Although the proportion of particles colliding with enough energy to overcome the activation enthalpy will be the same, the number of these collisions will increase.

2 (a) II (b) II

The effect of temperature on rate

Page 55

1 The activation enthalpy.

2, 3, 4

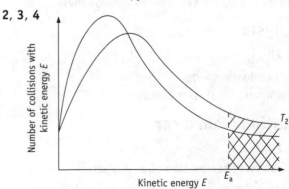

How do catalysts work?

Page 56

1 The MnO_2 provides an alternative pathway for the reaction with a lower activation enthalpy. This means more particles collide with the required energy, more successful collisions occur and the rate of reaction is faster.

The Polymer Revolution (PR)

Alkenes

Page 59

1 $CH_3CH_2CH_2CH=CH_2$

2 ⋀⋀

3

```
   ┌ Cl  H ┐
   │  │   │ │
 ──┤  C───C  ├──
   │  │   │ │
   └  H  Cl ┘ₙ
```

Forces between molecules: temporary and permanent dipoles

Page 61

1 Xenon. An atom of xenon has more electrons than an atom of krypton so the instantaneous dipole–induced dipole interactions are larger.

2 In branched chain hydrocarbons the molecules are unable to lie close together so the intermolecular forces are lower than in close-packed straight chain hydrocarbons. The higher intermolecular forces in the straight chain molecules means it is more difficult to pull them apart. More energy is needed, so the boiling points are higher for the straight chain hydrocarbons.

Structure and properties of polymers

Page 63

1 Thermoset plastics have cross-linking between the chains so will not melt. Thermoplastics have no cross-linking so when they are heated they can be re-formed and they will hold that new shape on cooling.

2 Reduce branching, increase chain length, introduce polar side groups.

3 Small unsaturated molecules (monomers) join together to produce a long chain saturated molecule (polymer). No other product is formed. The mechanism is free radical.

Forces between molecules: hydrogen bonding

Page 65

1 F, O, N

2 It can form hydrogen bonds with the water.

3 The H_2O arranges itself into an open lattice in order to maximise hydrogen bonding.

The –OH group in alcohols

Page 67

1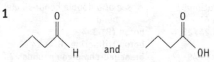

2 Mixture would stay orange. 2-Methylpropan-2-ol is a tertiary alcohol so does not undergo oxidation under these conditions.

3 Hex-1-ene

Geometric isomerism

Page 68

1

```
   Br      Br          Br       H
     \    /              \     /
      C=C                 C=C
     /    \              /     \
    H      H            H       Br
```

cis-1,2-dibromoethene   trans-1,2-dibromoethene

Index